ROCKET GIRL

GEORGE D.
MORGAN

ROCKET GIRL

The Story of Mary Sherman Morgan,

AMERICA'S FIRST FEMALE ROCKET SCIENTIST

Prometheus Books

59 John Glenn Drive
Amherst, New York 14228–2119

Published 2013 by Prometheus Books

Rocket Girl: The Story of Mary Sherman Morgan, America's First Female Rocket Scientist.
Copyright © 2013 by George D. Morgan. All rights reserved. No part of this publication
may be reproduced, stored in a retrieval system, or transmitted in any form or by any
means, digital, electronic, mechanical, photocopying, recording, or otherwise, or con-
veyed via the Internet or a website without prior written permission of the publisher,
except in the case of brief quotations embodied in critical articles and reviews.

Prometheus Books recognizes the following trademarks, registered trademarks,
and service marks mentioned within the text: Apple®, Boeing™, Boy Scouts of America®,
Chevrolet®, Chrysler®, Coca-Cola®, Craigslist®, Facebook®, Ford®, Formica®,
Gallup℠, Harley-Davidson®, iCloud®, IMDb®, Jeep®, John Deere®, Kodak®, Lifetime®,
Lucky®, Mercedes®, Mustang®, Pratt & Whitney®, Rocketdyne®, Trivial Pursuit®,
Volkswagen®, Volkswagen Bug®.

The Internet addresses listed in the text were accurate at the time of publication. The inclusion of a website does not indicate an endorsement by the author or by
Prometheus Books, and Prometheus Books does not guarantee the accuracy of the information presented at these sites.

Cover design by Nicole Sommer-Lecht
Cover photo of Mary Sherman Morgan courtesy of G. Richard Morgan
Cover image of the Redstone/Jupiter C rocket with *Explorer 1* satellite courtesy of NASA

Inquiries should be addressed to
Prometheus Books
59 John Glenn Drive
Amherst, New York 14228–2119
VOICE: 716–691–0133 • FAX: 716–691–0137
WWW.PROMETHEUSBOOKS.COM

17   5 4 3 2

Library of Congress Cataloging-in-Publication Data

Morgan, George D.
    Rocket girl : the story of Mary Sherman Morgan, America's first female rocket sci-
entist / George D. Morgan.
        pages cm
    Includes bibliographical references and index.
    ISBN 978-1-61614-739-6 (pbk.)
    ISBN 978-1-61614-740-2 (ebook)
    1. Morgan, Mary Sherman, 1921-2004. 2. Rocketry—United States—
Biography. 3. Women scientists—United States—Biography. 4. Rocketry— United
States—History—20th century. I. Title.

TL781.85.M67M67 2013
509.2—dc23
[B]

                                                        2013010090

Printed in the United States of America

CONTENTS

6 CONTENTS

FOREWORD

I remember, at three years old, watching Apollo 15 and seeing the astronauts drive on a non-terrestrial body for the first time. I didn't know their names at the time, but I sensed the importance of the moment and the specialness of these people. I confidently announced that I was going to be an astronaut and go into space. My parents didn't laugh. My mother didn't tell me I should think about being a teacher or a nurse or a secretary instead. My engineer father never doubted that I could follow in his footsteps, nor saw any reason why I shouldn't. They just said "Okay." I have had that sort of support and encouragement throughout my life—from my parents, my teachers, my mentors, and my friends. I majored in physics at a technical college and completed a PhD in robotics. Today, though not an astronaut, I have gone into space as an engineer at the Jet Propulsion Laboratory.

But this world of open doors in which I was raised is not the world Mary Sherman Morgan faced. In that world, women were not considered capable of technical work. The world that I took for granted was born slowly and painfully through the perseverance, dedication, and mostly unknown efforts of exceptional women like Mary. The growing pains began in earnest during World War II. Technological and scientific advancement was essential for Western survival, but men who would otherwise hold these jobs were suddenly overseas. The vacuum was filled by those left behind—women like Mary, who insisted on taking every science class in school despite all objections. For the first time, women were given opportunities to apply these skills to something other than teaching, even if, like Mary, they didn't have advanced education. And the country had a vested interest in their success. After the war, women were expected to simply step aside and slip back into the home or the classroom, and most had no choice but to do so. Some remained in subordinate jobs due to economic necessity. But some, like Mary, remained to satisfy newly found professional confidence and satisfaction, and because they now had rare expertise in new fields. As her son says, she was "determined not to be pigeon-holed" and she found a new place to apply her expertise and commitment in aviation and rocketry.

I first heard of Mary Sherman Morgan when I saw the premiere of the play *Rocket Girl* at Caltech in 2008. Parts of the story her son George told were familiar to me from history—skepticism of her abilities, lack of recognition, low pay, and being the only woman in her environment. But what is unique about Mary is that she let her achievements speak for her. She didn't fight a public battle about whether women could or should do the work—she just *did* the work. Perhaps she thought a woman's work would not be trusted. Perhaps she knew that attitudes can't be changed by force or by confrontation. She slowly and subtly changed the attitudes of the men (and women) around her by simply, quietly, being herself.

Unlike Mary, I never faced the still predominantly male world of engineering as the only woman. I did, on occasion, encounter those who believed women were not as capable, but I was able to dismiss them and not let their doubts become mine. While workforce equality has some way to go, there are few "firsts" left for women to conquer. By reading Mary's story, we are reminded how far we have already come and, more importantly perhaps, we may learn how to successfully complete the journey. When George asked me to write this foreword, I wondered what Mary, who shunned public and professional recognition, would think. I hope that she would appreciate the need to celebrate those like her who persistently aspire to, and do, achieve something extraordinary—not extraordinary for a woman, but extraordinary unqualified. And in this new world Mary helped create, that is something to which we *all* can aspire, while simply, though not necessarily, quietly being ourselves.

Ashley Stroupe

Ashley Stroupe, PhD, is a robotics engineer at the Jet Propulsion Laboratory in Pasadena, California. She performs research and supports operations of spacecraft. In 2005 she became the first woman to remotely drive a vehicle on another planetary body. She is currently one of a handful of "pilots" trained to drive the Mars rovers.

1.
THIS IS A STORY

This is a story about a mother who never talked to her children. This is a story about a wife who rarely talked to her husband, though they were married for fifty-three years. This is a story of a woman who desperately wanted happiness but could never summon the strength to reach for it. This is a story of a woman who had a family that loved her, but who struggled to love them in return. This is a story about a woman whom people admired but could never get close to. This is a story of a woman who harbored many secrets and lived in daily fear that those secrets would one day be revealed. This is the story of a woman who took those secrets to her grave. This is a story about America's first female rocket scientist.

This is a story about my mother.

Mary Sherman was born on November 4, 1921, on a small farm in a remote corner of North Dakota. There is no record of who was present that day, as the Shermans were never great record keepers. On August 4, 2004—eighty-two years later—Mary was admitted to the emergency room of West Hills Hospital in West Hills, California, with chest pains. Her husband (my father), G. Richard Morgan, was by her side. An hour later, Mary was dead. There is no record of who else except my father was present, as the Morgans have never been great record keepers. They wheeled her body out of the room, and a nurse collected the possessions she left behind: a handful of home-sewn clothing, fifty feet of clear plastic oxygen tubing, and a plastic bag overflowing with exotic medications.

That afternoon, my father began calling his children, which

included my brother, Stephen (a long-haul trucker), my sister Monica (a draftsman living in Oregon), and my sister Karen (a government health worker in nearby Orange County). He called me last and asked that I write Mary's obituary for the *Los Angeles Times*. I told him I would be honored.

I did not expect any challenges writing her obit. Even though my mother always refused to discuss her 1950s top secret Cold War work with any of her four children, we had learned quite a few things from eavesdropping on little snippets of conversation between our parents and their friends. We knew, for instance, that our mother had been a rocket scientist, that her work included designing new and exotic rocket propellants, and that she had made several historical achievements that helped usher in the Space Age. Taking this obituary assignment most seriously, I interviewed my father, found out a number of things I never knew about my mother, wrote the obit, and submitted it to the *Times*.

I assumed that getting her obituary published would be a slam dunk, given that my mother was the inventor of hydyne—the rocket propellant that boosted America's first satellite, *Explorer 1*, into orbit. Her invention had helped rescue America's tarnished reputation in the wake of Russia's launch of *Sputnik 1* and *2*. It was a significant milestone in the history of America's space program, especially since she had been the only woman out of nine hundred engineers, and she didn't even have a college degree.

To my surprise, however, the *Times* refused to print the obit. The reason, they said, was that my mother's life "could not be independently verified." They said that they had checked the claims in my obit article and could not verify any of them.

They could not verify *any* of them!

That was when I realized my mother's numerous accomplishments in the fields of rocketry and aerospace were already turning to dust and were in danger of being lost to history forever. Apparently no one at North American Aviation had been very good at keeping records either, because two years later a former NAA engineer, Robert S. Kraemer, would write a book about the company. He would do it because no one else could. And the reason no one else could write it? According to Kraemer, "The professional historians said there was not enough preserved documentation for them to write a proper history."[1]

After seven years of working on this project, I can tell you this: Robert Kraemer and those "professional historians" were absolutely correct. Historical record keeping in the non-aircraft portions of aerospace has been abysmal.

My mother had been a devout Catholic most of her life, so the funeral service was held at St. John Eudes Catholic Church in Chatsworth, California—just a few miles from the home Mary lived in for almost forty years. Despite the fact I wasn't a Catholic, the family agreed that I should be the one to deliver the eulogy. It was pretty normal stuff as eulogies go—lots of talk about Jesus, the resurrection, heaven. Blah, blah. After the services, all the attendees—about fifty people—sat down for a sunny, outdoor lunch in the church courtyard.

Lining up at the buffet table, I dished up a plate of food and went looking for an available seat. I noticed that a number of my mother's former coworkers were gravitating to a table set off from the rest of the group.

My mother had chosen to share very little information about her life as an aerospace engineer. She claimed that her security clearance forbade her from doing so, but we always suspected there was a lot more to it. Long after she retired, and most of her work had been declassified, she continued to enforce those rules of secrecy on herself. She had seen too many of her friends punished over the years for the smallest of infractions. Most of her top secret work had been performed in the 1950s—the McCarthy era—and people were afraid. But long after Senator McCarthy was dead, his ghost continued to haunt my mother's soul. So when I noticed the NAA engineering group congregating at a single table and talking over the "good old days," I knew where I had to set my plate. I was eager to hear a few war stories from these men (and they were all men). I took a seat at their table, took a bite out of my sandwich, and quietly listened.

Not more than two minutes went by, however, before I felt a stern tap-tapping on my right hand. I turned to see the face of a very elderly gentleman sitting across the table, his face wrinkled and folded, like those canines that win the "ugliest dog" contests. He stopped tapping my hand, using his bony index finger to point straight at my nose. He spoke.

"You need to listen to me, young man."

"Yes, sir."

"My name is Walter. I knew your mother. I worked with her. I'm going to tell you something about her you probably don't know. Listen carefully."

"I'm listening." That was the truth.

He looked left and right, as if checking for FBI surveillance, then stared though my body like it was made of glass.

"In 1957, your mother single-handedly saved America's space program," he said, "and nobody knows about it but a handful of old men."[2]

"Hm. Okay."

"You need to tell her story," he said. "You need to let people know the truth. Don't let her die nameless."

That's when I remembered: the monthly bridge games.

As a young boy, some of my earliest memories were the bridge tournaments hosted in our Reseda, California, home by my parents. At that time, they were both working for North American Aviation— the forerunner of Rocketdyne. It was the place where they had met. Marrying my mother was no small competitive feat for my father, since she was, as I've said, the only woman out of nine hundred engineers. Once a month, about a dozen of those engineers and their spouses would gather at our home to socialize around the card tables. I would walk amongst those tables, small and anonymous, listening to phrases such as, "Two spades," "Three hearts," and "We had a fire on the test stand today." Walter, I now remembered, was one of those bridge-playing engineers.

As he began eating his lunch, I told Walter how I had been unable to convince the *Los Angeles Times* to publish her obituary. He nodded understandingly.

"To get a large city paper to publish an obit the deceased has to be famous—something your mother was not."

Despite being a pioneer in the all-male world of aerospace engineering, and despite a long résumé of important and historical accomplishments, Mary had worked hard at not being famous.

"I do not want to see my name in print. You will not write articles about me—not while I am alive."

Those were my mother's words to me just after her eightieth birthday when I had the temerity to suggest it would be a good idea if I wrote a magazine article about her historic contributions to American rocketry. When I pressed the issue further she became belligerent, even angry. This was a woman who cared nothing for notoriety, a true anachronism in today's celebrity-obsessed culture. Mary Sherman Morgan was a woman who shunned publicity and valued her privacy more than life itself. She hated celebrity and detested those who sought after it. To put it another way, she was the exact opposite of that avid publicity hound Wernher von Braun.

Humility, however, has a downside; its practitioners can be lost to history, no matter how great their accomplishments. My final phone conversation with the editor of the *Los Angeles Times* obituary department grew heated as she continued to refuse to publish my mother's obit. When the argument reached a red-faced crescendo, and she continued to be obstinate, I threatened to take some kind of action.

She replied, "What are you going to do, Mr. Morgan—sue us?"

"Oh, no. I'm going to do something much worse than sue you," I said. "I'm going to write a play."

I hung up the phone and immediately opened my laptop. Through the magic of theater, I decided, I would accomplish what history, the army, NASA, the media, and my own mother had refused to do: I would write a play and use it to bring Mary Sherman Morgan's accomplishments into the light of day.

This self-imposed assignment quickly turned into a journey—a journey that would take me to many places as I played detective, tracking down the small number of former coworkers who were still alive. They were all retired, of course—some for decades. When I told them about what I was doing, they were unanimous in their desire to help.

You need to tell her story. You need to let people know the truth. Don't let her die nameless.

In November 2008, the play *Rocket Girl* opened at Caltech's 400-seat Ramo Auditorium, playing to large, enthusiastic audiences. At the end of each performance, mothers and their daughters would come up to me and tell me how inspiring the play had been for them. By the time the curtain closed on the night of our last performance, hundreds of websites across the Internet spectrum were reporting, talking, and blogging about Mary Sherman Morgan. Mary became the subject of a history fair exhibit by a high-school student in Maryland (the young girl won the state championship with it), a docent at Cape Canaveral began incorporating my mother's story into the official tour guide spiel, and theaters around the country began inquiring about producing the play. And even though the *Los Angeles Times* still refused to publish her obituary, Mary's name became known to millions almost overnight.

That could have been the end of the story, and in fact I did expect the responsibility to my mother's legacy to end with the closing of the play. I had done what Walter had instructed me to do—I had told my mother's story; I had told people the truth. I had made sure she did not die nameless.

This is a story that should have ended right there. However, the production of the play, and the subsequent media attention it garnered, triggered a series of events I could never have foreseen. With the storm of attention that followed the play, we would all soon discover why our mother had spent her entire adult life refusing to talk about herself or her past. Unbeknownst to me or my siblings, Mary Sherman Morgan had lived her life hiding a number of secrets. The media attention that followed the play became the earthquake that rocked the foundation of those secrets and forced them out into the open.

This is a story that should have ended in November 2008. Instead, it was just the beginning.

2.
PRAIRIE GIRL

Unsympathetic bankers, a hardscrabble land, and a growing season shorter than a John Deere brake shoe had for years combined themselves into a conspiracy of execution. It was here in a far corner of North Dakota the dreams of European immigrants came to die.

Undulating plains of grass extended in all directions to the horizon. To the south, a retreating gray mass of falling mist from some distant storm roared onward. Cutting the grassy plain in two, like a finger run through fresh paint, was a road. In summer the road was hard dirt. But this was late November, and after two days of rain it had become greased with the slimy mud of Dakota farm country. Pitted with water-filled ruts, and combed with gouges from old tractors, car tires were known to disappear into this muck at random. Fortunately, the driver of the 1930 Chrysler had been here before and knew where the soft spots were.

The Sherman farm soon came into view, its half-finished little spit of a house hardly bigger than a two-horse barn. It sat there atop its meager ten-acre plot, a fading brown memory waiting for time to finish its work. The bone-white paint that once adorned the structure had long since been blasted away by the Canadian winters, their winds indifferent to man's imaginary borders, blowing ever southward through the flatlands like giant walls of sandpaper.

As the Chrysler approached, its passenger could detect a new sound wafting up along the air currents. It was the rhythmic creak of metal hinges, endlessly complaining about the farmer's refusal to oil them. Betty Manning, a supervisor with Williams County Social Services, turned to look out the rear window, making certain the horse trailer was still attached.

The Sherman farm had once been encircled by a wooden fence, but it had collapsed long ago. Only the gate remained—a lone sentry protecting nothing from nothing. The Chrysler arrived and the driver, Sheriff Knowles, pushed the gate open with a gentle nudge from the car's chrome bumper. Ten feet from the front porch, Sheriff Knowles cut the engine. He looked out his window to examine the front tire, now sunk a full hand-width into the wet sludge.

"Thirty miles from town is no place to get stuck."

The "town" was Ray, a rail stop along State Highway 2 populated by three hundred hardy souls. (Seventy years later the population would grow to four hundred.) Small by any standard, its central location amongst thousands of square miles of farmland would one day inspire its residents to refer to their town as the "Grain Palace City."

Betty opened her door and stepped out. "Rain or no we have a job to do." She headed toward the farm house porch. It took only two steps for the mud to swallow one of her shoes.

"I'll get it." Sheriff Knowles retrieved the shoe, then helped her to the porch.

Betty was about to knock when she was distracted by a movement through the window. A young girl had been watching them, stepping out of view once she realized she might be spotted.

"Mary? Is that you in there?"

But there was no answer.

"Do you have the papers?" the sheriff asked.

Betty nodded, pulling them from a deep pocket in her dress. "Do you have handcuffs?"

He indicated where they were clipped to his belt.

"If he's been drinking it might get nasty." She raised her fist to the split-pine door and knocked loudly.

"Mister Sherman," she called out. "It's Betty Manning from Social Services."

The sheriff leaned to one side to peek through the window. "He's sitting in a chair. Asleep—or passed out."

Betty knocked again. "Mister Sherman—please open the door."

They waited, but when the door remained shut the sheriff reached out, turned the knob, and pushed the door open.

"Mike?"

Michael Sherman was hunched forward in a wooden chair, drunk and still. Around the chair, like a protective circle of wagons, lay a ring of empty wine bottles. At the far side of the room, eight-year-old Mary stood alone, pale and quiet.

"My dad's asleep."

Sheriff Knowles shook the man's shoulder. "Mike—wake up. Come on, Mike."

A single eye opened—then the other. He mechanically turned and tilted his head to address his rouster.

"Jeff? Whatta you doin' here?"

"I'm afraid I'm going to arrest you if you don't start obeying the law."

Mike noticed there was a second person in the room. He turned slightly to see who it was.

"Oh. You."

Betty stepped forward. "Mister Sherman, your daughter still is not attending school. She is eight years old—more than two years behind her peers. If you do not enroll her today you will be arrested."

"I need her to clean the creamer."

"Mary needs to be in school, sir."

"This is a farm; we have chores."

"This is country; we have laws."

Michael turned his head and spat at an insect that had crept in through the floorboards. "How's she gonna get across the river?"

"We've already been over this, Mister Sherman. The State of North Dakota will provide a horse she can use to get across the river to the schoolhouse."

The Sheriff crouched down to talk to Michael face-to-face.

"Mike—Mary needs to be in school."

"They took everybody. All my kids; Amy, the boys; everyone's gone. To school." He pronounced the word *school* with a deep drawl of disdain.

"I need somebody to help me, Jeff. Elaine's too young to do much. All I got is Mary."

A young girl entered from a back bedroom and stood next to Mary. The sheriff addressed them.

"Girls—where's your mom?"

Mike answered for them. "She's visiting her sister."

The sheriff reached out to Betty, who handed him a sheaf of papers. "It's not just the law, Mike. It's what's right. It's what's right for your daughter."

Michael took the papers. The sheriff removed a pen from his crisp uniform pocket and offered it up. As his eyes became adjusted to the dark room, Sheriff Knowles could better see the worn-out man sitting before him; the whiskery face and reddened skin, the unkempt hair, the half-open eyes peering through clouds of glaucoma and alcohol.

Michael managed to find the pen, took it, and affixed his signature.

"Now who's gonna clean the creamer?"

Betty took the signed forms. "Why don't you give it a try, Mister Sherman."

She stepped farther into the room so Mary—hiding behind a small table—could see her.

"Mary. My name is Mrs. Manning. Do you remember me?"

The young girl nodded quietly, and Betty stepped closer.

"Would you like to go to school? Right now? There's lots of other kids waiting to meet you."

Mary glanced at her father to see his reaction, then turned back and nodded her head.

"Come on," said Betty, offering her hand. "I have a surprise for you."

Mary stepped forward and entered a ray of sunlight that poured in through a cracked window. She wore a threadbare print dress, but no shoes. Her face was sullied, her hands filthy. Her hair was scraggily and oily and seemed to have attracted two flies in orbit. Sinfully near-sighted, her eyelids were affixed in a near-permanent squint.

Mary took the woman's outstretched hand and followed her to the door. She stopped next to her father.

"Daddy? May I go outside?"

Michael said nothing for several moments. He finally broke the suspense with a single word.

"Go."

"Come on," whispered Betty.

As the sheriff signaled he would stay with Michael, Betty led the young girl outdoors, but not before giving her sister Elaine a sideways glance. Betty knew it was just a matter of time before she would be back for the little one.

Most of the storm clouds had moved on, leaving a brilliant midday sun to brighten the Dakota prairie. A slight breeze blew, creating waves along the tops of the grasses and pulling with it the dreams of young girls.

"Do you like horses?"

Mary nodded.

"Would you like one for your very own?" Betty walked to the back of the trailer with the girl in tow. Mary could hear the animal's huge lungs breathing in and out as the woman opened the rear trailer door.

"Come on, boy." She pulled out a short wooden ramp and led the horse down to the muddy ground.

"What's his name?"

"I don't know. Why don't you give him a name?"

Mary thought a while, looking around the farm as if for some sign of inspiration.

"I'm going to name him Star."

"Star?"

"Every night I like to come out here and look up at the planets and the stars. I like to watch them move across the sky. I'm going to name him Star."

"Then that's his name."

"Am I going to ride him to school?"

"Not for a while, honey. We need to get you some riding lessons first. For the next few days Sheriff Knowles will be driving you to school. A horse is a big responsibility. Once you know how to ride him—and take care of him—Star will be all yours."

Betty tied the horse up to the lonely gate. The sheriff came out of the house, removed two bags of feed from the trailer, and set them on the porch. Then he opened the rear door of the Chrysler.

"Ready to meet your new classmates?"

They got into the car and the sheriff started the engine. He pulled the car around and headed back toward the highway.

The two lane road—Highway 2 on the map—followed every rise, descent, ridge, and rill of the landscape, its engineers having made no attempt to straighten out Mother Nature's work. Betty glanced back to see how her charge was handling the ride.

"When we get to the school we'll clean you up a bit. Before we introduce you. Okay?"

Again Mary quietly nodded.

Betty turned and watched the road spool beneath the car, like some oversize conveyor belt. She had no way of knowing the historic effect the young girl behind her would have. Mary was just another neglected kid who needed to get out of the farmhouse and into the schoolhouse. Betty handled at least two dozen such cases every year—nothing special about this one.

—◦\/\/\◦—

The "War to End All Wars"[1] had been over for more than a decade. Yes, the Great Depression was in full force, but Ray, North Dakota, was hardly Detroit or New York; the woes of Wall Street were barely felt in the northern frontier. Here on the Dakota prairie, economic depressions neither arrived nor left. At least the world was at peace. Yet even now there were rumblings in Europe—the humiliation of Germany having been a gag in the throat of its people. Soon the land of Prussia would be home to great factories building secret and mysterious weapons—weapons that would spread fear to every country. One of those weapons would evolve into the seed of a future space race between the world's two most powerful countries.

Mrs. Manning had no way of knowing, of course, that the scrawny and scraggily eight year-old girl seated behind her would play a major role in that race. She had no way of knowing that a second world war was in the offing—that the unwashed little urchin she was now escorting to school would one day contribute to the war effort as a chemist in a weapons factory. Or that she would play a pivotal role in the coming space race. Or that she would one day rise to become America's first female rocket scientist.

Or that she would become a champion bridge player.

Betty Manning could know none of this, all of it being too far ahead to see. Little Mary Sherman was just a child, a face, a name. A name like Jennifer or Susan or Elizabeth or Star. A name on a very long list of names.

The one-room schoolhouse came into view over a rise. Betty Manning—dedicated social service worker with Williams County, North Dakota—opened her briefcase and pulled out her next assignment.

3.

THE RAKETENFLUGPLATZ

"For my confirmation I didn't get a watch and my first pair of pants, like most Lutheran boys. I got a telescope."

—Wernher von Braun[1]

At the same moment that Betty Manning was marching the urchin farm girl toward Ray's one-room schoolhouse, Captain Heinrich Strugholdt, Berlin's chief of police, was doing a little marching of his own. Moments before, a homemade steel rocket had slammed into the Berlin North Police Station,[2] puncturing a four-inch hole in the station's roof and filling every cubic centimeter of the building's interior with noxious fumes. The three-foot-long rocket had come to rest atop the desk of one Officer Ernst Ritter, who had been so terrified by the experience he had to go home and change his underwear.

Captain Strugholdt clutched the rocket firmly in his right hand as he negotiated the twelve steps outside the main entrance. There was no doubt where the rocket had come from, or who owned it. The small cadre of young college students was well known around the city for their rocket experiments at an abandoned World War I ammunition dump just north of the city. Consisting mostly of bleak, open fields of grass and swamps, the surplus property was owned jointly by the city of Berlin and the German Defense Ministry. The land and its few scattered, dilapidated buildings had been lent to the young rocketeers thanks to the space-dreaming, fast-talking, entrepreneurial salesmanship of one Rudolf Nebel,[3] a college student with dreams of building a rocket to fly to the moon. Captain Strugholdt had personally visited the area several times to watch the launchings and inspect the goings-on. None of the students' rockets ever seemed to go high enough or far enough to reach occupied buildings—the closest being more than a kilometer away. Clearly, however, the work of these enthusiastic rocket experimenters had progressed considerably since his last visit.

As Captain Strugholdt approached the city boundary, the four-story office buildings gave way to two-story hotels and apartments, which then yielded to small shops and markets. Passing Lars Michel's shoe store and a few small residential cottages, the road narrowed and the vista opened, revealing the raw, undeveloped fields and brilliant green grass of the ammunition dump's periphery. It was this field that the youthful rocket experimenters had begun referring to as their *Raketenflugplatz* (Rocketport).

At the point where the road narrowed, the pavement made a wide bend to the left, then continued straight as it ribboned its way northward toward Germany's border with Denmark, a three-hour drive away. It was at this moment, as Strugholdt was entering the grassy field, that he encountered the students, approaching on foot and searching the tall grass for their errant missile. At the sight of their rocket, the boys rejoiced.

"Captain Strugholdt—you found our rocket!"

"Oh, yes," he replied. "I found it."

The captain had never bothered to investigate this group closely, a decision he now regretted. The local residents never complained and, in fact, were often seen joining the experimenters tourist-like during their monthly launches. Strugholdt knew none of the names of these boys, who they were, what universities they attended, or who their families were. Today, however, someone had to be held responsible.

"Which of you is the leader?" the captain demanded.

The boys exchanged puzzled looks. None of them had ever been technically "in charge"—they were a headless ragtag, a rudderless ship, a group led by the nose of enthusiasm rather than some outstanding personality. Today, that would change.

"I said, who is in charge!?"

One by one, the youthful rocketeers turned to look at the muscular, attractive, blue-eyed blond boy walking out of a tall strand of grass. The boy brushed dew and thistles from his pants as he approached. He was dressed considerably better than the others, a sure indication of his family's wealth. The boys pointed to him.

"He is."

"What is your name?" the captain demanded.

"Wernher," the boy said. "Wernher von Braun."

"You and your friends are not welcome here anymore," said the captain. "Go home. Tell your parents they owe me two thousand marks for the damage you caused to the police station."

With that, the captain turned and marched back toward the city, the rocket still firmly in his grasp. As the boys quietly watched him go, smiles began to spread across their faces. If, in fact, their rocket had managed to reach the police station, then that could mean only one thing: their many hours of work and research and testing were paying off—their rockets were becoming more powerful and sophisticated, reaching greater heights and longer ranges. They congratulated each other on their success, went their separate ways, and never told their parents.

4.

THE ABC'S OF MILKING COWS

Mary gently pushed open the door to their home. Tightly holding her first homework assignment with both hands, she passed the threshold and entered the dim house. There was a *clunk*, and she looked down to see her foot had knocked against an empty wine bottle. Mary looked around but saw no one. She stepped forward.

"Ya still have to do your chores."

Her father's voice startled her. Now she could see him, his form hidden in the moody shadows of the late afternoon. He was sitting in a chair, staring into space.

"Chores never end on a farm."

"But I have homework."

"Don't matter. Ya wanna go to school? Fine. Ya still have to do a full day of chores—just like everybody else."

"But I won't have time for my homework."

"Just like everybody else."

The barn door was missing a hinge and Mary had to lift it slightly in order to close it. The cow's name was Irma, and as Mary led her slowly toward the milking post, she was careful not to soil her shoes. They were scruffy and well used, but they were hers. Hand-me-downs from a cousin, who had inherited them from some other cousin, they were perfect. It was Aunt Aida who had personally instructed her parents to get Mary some good walking shoes, and they had agreed. In all the world, Mary owned four personal possessions: two dresses, a corncob doll from Aunt Aida, and now the shoes. Were she to be buried with them, archeologists a thousand years hence would make note of how poor the family must have been.

Mary heard the sound of the truck's engine starting and turned to see Vernon and Michael climb over the gates and onto the flatbed. As her father circled the truck around the yard once, Vernon whistled loudly, and their three dogs, each of them named Rover, jumped into the moving truck. As the truck drove away, Mary's brothers made ugly faces at her. She turned away, grabbing the metal bucket with one hand, the rickety egg crate with the other.

"Hi, Irma. Ya got some milk for me tonight?"

The leather belt was already in the crate. Mary removed it, then wrapped the belt around Irma's hind legs to restrain her. She pulled the belt tight, then set the buckle. Irma huffed a breath of air through her downy, white nostrils, giving a subtle grunt as Mary took her seat on the crate. The wobbly wooden box settled several inches into the soil—a wet, slurry of mud, manure, and muck, all mixed with the downstream offal of the butcher block.

"Actually, I should say *you*. Mrs. Bowman—that's my teacher—says I shouldn't say *ya* because *ya* is not a word. You should say *you*, she says. Never say *ya*—only say *you*. Only the ignorant use words that don't exist. That's what she says. So let me try that again. *You* got some milk for me tonight?"

Mary placed the bucket beneath the cow, then ran her hand tenderly along the soft skin of the udder, gently massaging and kneading the bulging sac. This was a technique her father had taught her. He called it "letting the milk down."

"I'm starting to learn the alphabet. The other kids know it already, and some of 'em make fun of me. At least I know English. Most of the other kids, their families are from Denmark and Norway. None of them knew English when they got here, but now they talk pretty good. Mrs. Bowman says they all seem to pick it up real quick. Danish in September, English in December; that's what she calls it. Anyway, the alphabet. Would you like to learn it, too? I'll teach you."

Mary placed all five fingers of each hand around the front teats and began to draw Irma's milk into the metal pail. She saw movement off to her left and turned to find her oldest brother Clarence walking toward the farmhouse. The high school was an easy two-mile walk down Highway 2—no river crossings required.

"First, there's the letter *A*. It has two sticks that form a point—like the roof of a house—with a little stick that connects them in the middle. *A* is what's known as a vowel. Got that?"

Irma had no reply.

"Then comes the stick with the two half circles. That's a *B*. Next we have a circle with a piece cut out of it—that's a *C*. *A-B-C*. Mrs. Bowman says the alphabet is sometimes called the ABC's. Whatta ya think about that?"

Irma lowered her head and searched for a fresh stalk of grass, but there was none.

"Mrs. Bowman taught me how to make all the sounds of the alphabet. Did you know that *C* is pronounced just like the beginning of the word *cow*? It's the first letter in the word *cat*, which means it's probably the first letter in the word *cow*, too. That's what I think."

Irma let out a low moan, turning her head to get a better view of her loquacious milker.

"Next comes *D*—a stick with a half circle. Then *E*—a sideways table with three legs. *E* is also a vowel. After *E* comes *F*—which is like *E* but with the bottom leg missing. *G*—which I found out is the first letter in *God*—is like a *C*, but different. The letter *H* is like *A*, but the legs are straight instead of pointy. *I* is just a line—that's easy to remember. *J* is kinda like a fishhook."

Mary paused in her lesson, easing her grip slightly on the teats.

"That was wrong. I meant to say, *J* is *kind of* like a fishhook. Kind of. Turns out *kinda* isn't even a word at all. Who woulda thunk?"

Mary looked around to make sure no one was approaching or watching, then removed a small slip of paper from a pocket in her dress.

"Hold on. I can't remember the rest." She unfolded the paper, revealing a series of crudely scrawled notes. "Ah yes—*K*. *K* is kind of hard to describe. Then *L*—like an *I* but with a foot. After *L* comes *M*, which happens to be the first letter of my name. N is like a sideways Z. We'll get to Z in a moment. O is just a circle, and P is like B but without the bottom half-circle thing. Q comes next. Mrs. Bowman describes Q as an O that's been stabbed. That's funny. R is like K, but rounded at the top."

Mary folded the paper and returned it to her dress.

"I think I know the rest."

There was a sound—a footstep. Mary turned around to see Clarence right behind her.

"Gimme some milk."

"No, Clarence. There won't be enough."

"Gimme some milk, Mary!"

"Dad says no!"

Mary screamed at him as Clarence reached in and grabbed the bucket. She tried to take it back, but he held it away at arm's length, pushing Mary down with his free hand.

"Give it back!"

He offered the bucket, then pulled it quickly away again, laughing as she vainly struggled to reach it.

"I'll tell dad!"

"Go ahead."

Clarence lifted the bucket and drank. Mary helplessly watched as he guzzled all of her work away. He finished, then wiped his mouth with his shirt sleeve.

"You tell him, and I'll hurt you." Clarence tossed the bucket in the mud, then turned and walked toward the tall, dried-out wheat stalks. "Keep your mouth shut, little schoolgirl."

Mary picked the empty bucket out of the sludge. Now she would have to clean it all over again. The well spout was nearby and she pumped water into the pail, using a cotton cloth to dry and sanitize its interior.

"Only good milk is clean milk."

Mary returned to Irma and the egg crate.

"Sorry. We have to start over. Anyway, S is easy to remember because it looks like a snake, and it's the letter that *snake* begins with. *T* is like an *I* with a flat hat. *U* is like a bowl. *V* is like an upside down *A* but with no stick in the middle. *W* is like two *V*'s stuck together. *X* is shaped like those two boards at the front gate that cross each other. *Y* is like a person standing up with their arms raised, and *Z* is like a sideways *N*. You got all that?"

Irma let out a long and forlorn moo, as if to affirm she had understood the lesson well.

"So that's the alphabet. The alphabet is very important; it's what words are made of. You use letters to make words, and words to make sentences, and sentences to make paragraphs. That's what books are—a whole lotta words, sentences, and paragraphs. Whatta ya think . . ."

Mary stopped to correct herself.

"Whatta *you* think about that, huh?"

Before Irma could respond, a clap of thunder rolled over the prairie and found its way to where she was sitting. Mary stood up and gazed past the grasslands toward the North Dakota horizon. A fresh Canadian storm was brewing slow and sure, its angry charcoal cloud layers leapfrogging one another, battling for position and marching steadily southward. They would arrive soon enough.

"Gonna have more rain, girl. Lots of it."

Mary sat back down and picked up the pace of her work, concerned about being caught milking in the rain. Not that she cared for herself, so much. When it rained, drops of water would wash over Irma's back, pick up the dust and dirt from her skin, and slop like grime into the bucket, contaminating the milk.

"Only good milk is clean milk," she said, briefly turning her head to check on the clouds one last time. "Only good milk is clean milk."

Irma agreed, and said as much.

A half hour later, Mary had a brimming bucket. As she toted it toward the creamer, her father and brothers returned in the truck, its cargo bed loaded with hundreds of pounds of lignite. With the temperature dropping seemingly every day, they would be burning lignite by the ton. Mary watched the men shovel the fuel into the storage bin.

Lignite had always fascinated her. How, she wondered, could a rock that was pulled from the ground actually burn? It was a puzzle, and she made a mental note to ask Mrs. Bowman about it.

Mary arrived at the creamer and raised the lip of the bucket to its rim. Tilting the bucket slowly, the raw milk poured into the machine. In a few moments, the bucket was empty.

After unbuckling Irma's hind legs, Mary led the cow back toward the barn for her late-afternoon feeding. As she walked, she quietly repeated the alphabet over and over and over.

5

I HAVE NO IDEA WHAT YOU'RE TALKING ABOUT

"I do not have OCD, OCD, OCD."

—WIDELY ATTRIBUTED TO EMILIE AUTUMN

I stare at my laptop, eyes drooping, breath slowing. According to the tiny digital clock in the bottom right-hand corner of the monitor, it is 1:32 a.m. I am exhausted. My wife and I have recently become foster parents and Ventura County has just placed a two-year-old boy with us. His name is Dalton, and he works overtime to find ways to keep us busy. As a result, most of my writing projects now have to be accomplished in the late evening, after everyone is in bed and things have quieted down.

The previous weekend I spent several hours interviewing Bill Webber, one of my mother's former coworkers. Bill had sat at the desk directly next to my mother's for several years, and now I'm sitting on a trove of great information. Finally, here is someone who knows where all the bodies are buried—far more so than even my father, who had worked in a different department than my mother. During our interview, Bill shared numerous anecdotes and historical details with me and, better still, shared my passion for getting my mother's story told.

Still, there is a gaping hole that cannot seem to be filled no matter whom I talk to.

For years before our mother's death, and long thereafter, my three siblings and I had often discussed what physical or mental ailment had afflicted our mother. Everyone knew about it, but no one talked about it. As I have mentioned, Mary Sherman Morgan never discussed anything personal to anyone, anytime, anywhere. And anything of a medical nature was at the top of the list of subjects over which she would rather die than discuss with her children. Yet it had always been obvious to us that something was wrong.

Two months before my brother Stephen was born, Mary decided to retire early from the aerospace business. It was December 1955. In a subsequent employment application two decades later she would write that her reason for leaving her prior employer was "to raise a family."[1] I wasn't even three yet, too young to remember much of anything. But over the next few years there were events and situations that began to accumulate in my memory. I remember our mother enjoyed playing four-handed bridge with herself—laying out four hands and strategizing how each hand would be played if it were hers (like a solo chess master playing both white and black). I remember walking home from school one day (yes, there was a time in this country when it was considered normal for very young children to walk to and from school by themselves) and arriving to a different atmosphere in the house. My mother was at the dining table with what had become her four props—a newspaper, a cup of coffee, a cigarette, and a deck of cards. She would read the paper, take a sip of coffee, take a drag on the cigarette, then repeat that action until the cigarette was gone and the coffee cup was empty. At some point she would remove the cards from their pack and begin shuffling them. And she would shuffle them, and shuffle them, and keep on shuffling them. Over and over and over. Eventually she would stop shuffling and play solitaire or solo four-handed bridge—but not until she had shuffled the deck a hundred times more than necessary. The detail I should stress here is that all this incessant shuffling was accompanied by anger—grinding teeth and furious mumbling. I was about six years old; I knew no other mothers to compare her to, so I thought it was normal behavior.

"George—who does Mom remind you of?"

It was my sister Monica, calling from Stayton, Oregon—the town she had hitchhiked to as a teenager thirty years before. She had gone there to discover herself. Instead, she discovered rain. Monica knew I was starting work on the play. I had told her to call me if she had any ideas or could remember any pertinent details I could use.

"Think," she said. "Who does mom remind you of?"

"What are you talking about?" Our mother had been dead for six months—she didn't remind me of anyone except herself.

"Have you seen that show on TV?" Monica asked.

"What show?"

"*Monk*. It's a TV show about an obsessive-compulsive detective. Have you seen it?"

In fact I had come across a few episodes. Not much of a TV watcher, nothing had clicked. What was she getting at?

"What are you getting at?"

"We've wondered for years what mom's problem was."

"Yeah, so?"

"I want you to imagine," she said. "Imagine the character of Monk without the funny—a Monk who's been taken over by the dark side of the Force."

I imagined it, and a light went on in my head.

"Yeah," I said. "A dramatic Monk instead of a funny Monk. Could be." I thanked her, and we hung up.

Was that her problem? Or perhaps I should say, was that her "condition"? Had Mary Morgan been obsessive-compulsive? I opened up a few websites and began studying the subject—a subject I knew nothing about. It didn't take long to figure out that this very possibly was the source of her many odd behaviors over the years—behaviors that at times were downright scary. As I read and studied, one other characteristic of the disorder also became obvious: there was absolutely nothing funny about it.

But here was the strangest thing of all: during my interview with my father, Bill Webber, and other former coworkers, I had asked about her odd repetitive behaviors. I wanted to know what it was like to work with someone like that. I wanted to know how it had affected her work, or her working relationship with her fellow engineers.

Their reply: "I have no idea what you're talking about."

Which was very strange. From my earliest youthful recollections, a day never went by that I did not witness her symptoms, if that's what they were. How could someone who worked next to her for years have missed them? I went back and re-interviewed my father on the subject; he still maintained the same reaction. This was a man who had been married to the woman for fifty-three years.

"I have no idea what you're talking about."

The house is cemetery quiet—the perfect environment to get some writing done. Have to get started. I need to get words on the page. I look at the digital clock in the lower right-hand corner of the monitor.

1:39 a.m.

All our lives we, the four Morgan kids, have been told the same your-mother-retired-to-raise-a-family story. But there are other things we have been told about her that have turned out not to be true. Does the retirement story fall into the category of Santa Claus and the Tooth Fairy? I'm beginning to wonder. I'm beginning to wonder if she was forced to retire from her top secret aerospace career because her OCD-like behaviors were becoming more pronounced and less self-controllable. People were fired all the time at North American Aviation for far less than that.

On this late evening as I struggle to get words on the page, I am fifty-two years old. Even now, many years later, I can still hear my mother grinding her teeth, mumbling to herself, and shuffling those cards.

I have no idea what you're talking about.

One day, completely out of the blue, I get an e-mail from someone claiming to be a friend of Irving Kanarek. I had been searching for Kanarek for years, but no one knew his whereabouts. In the '60s and '70s Irving Kanarek would become famous for being Charles Manson's defense attorney. But long before that career took off, Kanarek was a chemical engineer at North American Aviation and worked as my mother's immediate supervisor. Based on his birth year, I had come to the conclusion that he must have passed away. But the lady who has e-mailed me says Irving heard about my play at Caltech and he wants to talk to me. Her phone number is in the e-mail, so I call her.

"Irving Kanarek is still alive!?" I ask, incredulous.

"Yes," she says. "He just turned ninety, but he's alive and living in a motel in Costa Mesa."

Costa Mesa is a city in Orange County about forty miles south of Los Angeles. Since Irving was one of my mother's supervisors at North American, it's urgent I talk to him right away. The lady gives me his e-mail address and I fire him off a few sentences: *Mr. Kanarek, you don't know me, but you worked with my mother, Mary Sherman Morgan, at North American Aviation. I am writing a book about her life. If it wouldn't be too much trouble I would like to sit down with you for a couple of hours and ask you some questions.*

The next day he e-mails me his phone number and tells me to call him. I take a few hours preparing what I am going to say to the man who has become one of the most famous defense attorneys in history. I need the reception to be as clear as possible, so I eschew the cell phone and use a land line at my office. I want to keep a record of the conversation, so I hold the receiver in my left hand and hover a pen over a pad of paper with my right. My finger dials his number, and the line on the other end begins to ring. This was the conversation:

"Hello," said a voice.

"Is this Mr. Kanarek?"

"Yes. Who is this?"

"This is George Morgan."

"Did you get the bat?"

"The what?"

"The bat. Did you get the bat?"

"The bat? You mean, like a baseball bat?"

"Yeah. Did you get it?"

"I'm not sure . . ."

"The day you were born I went to visit you and your mom in the hospital. I brought you a present—a kid's baseball bat. Did you get it?"

"Ummm . . ."

"Ask your dad—he knows about it."

On the day of this conversation I am fifty-seven years old, but suddenly the image pops into my head. When I was very young, I had a small wooden bat, about half the size and weight of a regular bat. I always assumed my parents had bought it for me; they never said otherwise. It was the bat my father had me use when he first taught me how to hit a baseball. For a brief moment my mind tunes out the voice of Irving Kanarek and a long-ago memory of my father teaching me how to swing a bat pushes in to my conscious. I am about six years old, and we're in the backyard of our house in Reseda.

"Hold the bat like this, son. Two hands firmly around the neck. Bring it back over your shoulder, then swing, like this."

My dad and I both hold the bat, and he demonstrates how to swing it. Then he walks ten steps away and pitches me the ball, slow and underhand. I swing the bat, and miss. He pitches several more balls—I swing and miss every one. After a few more strikes, I throw the bat down in frustration.

"I can't do this!"

My father walks back and gets down on one knee.

"I'm going to tell you a little secret about hitting a baseball, okay?"

I nod my head.

"The hard part about hitting a baseball is you're trying to hit a round object with another round object. It's difficult. But the more you know about the science of physics, the easier it is. In my job I use a lot of physics, so I can give you information few other fathers know about—understand?"

I nod my head.

"I'm going to tell you a secret about hitting a baseball, and you have to promise not to ever tell anyone else. Okay?"

I nod my head.

My father looks left and right, making sure there's no one else around who will hear the secret. He holds up a ball, and whispers.

"Keep your eye on the ball."

He takes ten steps back, and pitches me the ball. I keep my eye on it, and swing the bat. The round bat connects with the round ball, and the little white orb goes sailing over our backyard fence. It hits the roof of the Velkers' home, knocking off one of their roof tiles. It bounces once and plops into their chimney. My dad turns and looks at me.

"Good. Now do it again."

Somewhere around the time I turned ten years old, the bat disappeared. God only knows what happened to it.

"Oh yeah, Mr. Kanarek. I got it. Thank you."

"You're welcome."

I set up an appointment to meet with Irving in Costa Mesa the following weekend. I thank him for his willingness to help with my project, and hang up the phone. For the rest of the day I can't help but have a recurring thought: The man who defended Charles Manson and the Onion Field killers gave me my very first birthday present.

6.

"MOTHER DOES NOT ABIDE PHOTOGRAPHY"

I n early 1935 Willy Ley, a young journalist, German rocketeer, and close friend of Wernher von Braun, convinced his supervisors in the Foreign Office to sponsor him on a journalism assignment to the United States. What Willy conveniently neglected to tell his supervisors was that he was secretly planning to permanently immigrate to America. The requisition for a "paid journalism assignment" was nothing more than a clever ruse designed to get Germany to pay for his escape.[1]

As a journalist, Willy Ley was more aware than most Germans of the sinister side of Adolf Hitler's rise to power.[2] He knew Hitler had murdered dozens of his enemies, and he suspected he would stop at nothing to become an all-powerful dictator. Ley was also aware of Hitler's quiet campaign to burn the Treaty of Versailles and remilitarize Germany. One of the most blatant violations of that campaign was the covert creation of a massive air force, what would come to be known as the Luftwaffe.[3] Willy Ley could smell another war coming and decided it was best to leave the country while the borders were still open. So it was that one cold morning in February Willy shared a beer with his good friend Wernher, exchanged a few warm good-bye hugs, shouldered his duffel bag, then boarded a steamer bound for New York. Neither of them could surmise the chain of events that, in a very short ten years, would result in Wernher following his friend across the Atlantic.[4]

For now, there was far too much action to keep Wernher firmly rooted in Germany—flying a rocket to the moon was his unyielding dream, and real progress on that dream was being made. The work of the German rocket boys was attracting attention—the kind of attention that writes checks. The previous December, the youthful von Braun crew had flown two of their most sophisticated rockets as a private demonstration for a group of army officers. Dubbed Max and Moritz, after a couple of kids in a popular German cartoon, the rockets broke all previous altitude records and made a significant impression on the visiting dignitaries.[5]

Since rockets had not been included in the Versailles Treaty as one of the verboten weapons, it left a major loophole the German army was only too happy to jump through.[6] Army generals and colonels began keeping a close eye on Wernher von Braun and his experimenting associates. Money had been flowing into the military like the Rhine River during spring runoff, and new technologies were given special funding priority. Any weapon that did not exist prior to June 28, 1919, was not a part of the Versailles Treaty and was therefore technically exempt. Von Braun and a small team of budding engineers were given a modest annual budget of 80,000 marks for research and development.

Such financial modesty did not last long, however, as it was soon overshadowed by the potent vanity of unlimited funding. In March 1936, the army's commander in chief, General Feiherr von Fritsch, wanted to see what his 80,000 marks was buying. He decided to pay a visit to the Kummersdorf—the remote location von Braun and his engineers had moved to after the police-station incident. What the general saw immediately impressed him. His exact words to von Braun were, "How much do you want?"[7]

After carloads of military money began arriving at the Kummersdorf, and hundreds of engineers, technicians, and clerks began to join the newfound payroll bandwagon, Wernher wrote a letter to his parents. In it he made a report of his recent successes, then he closed the letter with a simple observation, "As the Americans say, we have made it to the big time."

•—ᐧᐧᐧ—•

It was August 2007.

I took a felt pen and addressed the two 9 × 13 manila envelopes, one to the theatre department at the Massachusetts Institute of Technology (MIT), the other to the same department at the California Institute of Technology (Caltech) in Pasadena. These were the top two engineering schools in the country, both of which I had applied to (and rightfully been turned down by) during my senior year in high school back in 1971. Each envelope contained a copy of the just-completed play about my mother. As is common with many writers and their projects, the play was easier to write than the title. I struggled with a number of alternatives, and wished many times I had thought of *The Girl Who Played with Fire* before Stieg Larsson had used it. Every title candidate I considered was too esoteric, too weird, or just too simple. In the end I settled for simple, and titled the play *Rocket Girl*. Those two words, I decided, would not win any titling awards on Broadway, but they said it all.

The play had taken me a year to research and write, and now the time had come to place the messages in the bottles and toss them into the deep, wide ocean. I applied my DNA to the envelope seals, closed them, affixed the proper postage, then handed them over to the US Postal Service.

Based on a thousand past writing-submission experiences, I expected to wait at least six months for a reply.

Four days later, I received a letter with from some organization calling itself TACIT. I had no idea what TACIT was, but the return address was in Pasadena. I ripped open the envelope. Inside was a one-page letter from Shirley Marneus, the theatre director at Caltech. TACIT, it turned out, stood for Theatre Arts at the California Institute of Technology.

Shirley Marneus was a well-known institution, having directed thousands of Caltech actors over a thirty-five-year career. In her letter she expressed enthusiasm for bringing *Rocket Girl* to Caltech. Her exact words were, "Your play and this school are a match made in heaven." I mentioned how timing was important; the following year would be the fiftieth anniversary of the launch of America's first satellite, *Explorer 1*, and if at all possible I wanted the play to commemorate that event by having its premiere before the end of 2008. She assured me that was within their capabilities.

Thus began a series of e-mails and phone calls as Shirley and I discussed what would be required to mount the play's world premiere. Astonishingly, she never once asked me to rewrite or edit any part of the play—almost unheard-of in the world of new play production. This concerned me, since all experienced writers know that rewriting is a healthy and necessary part of the process. I let it get to me, of course, allowing my ego to expand like an inflated zeppelin (*The play must be perfect, blah, blah*). Shirley set a date five months away for opening night, and I could not have been happier.

Then reality set in.

A few weeks passed, then Shirley began calling me, groaning about "scheduling problems." Caltech had three venues suitable for staging a play, and apparently the school's faculty had all three of them booked well in advance for lectures, seminars, symposiums, and so on.

"We'll have to reschedule the production."

The play was pushed back a few months—several times. When the year 2007 came to a close, and we still did not have a firmed-up opening date, I began to worry. Shirley kept reassuring me, but as the sun rose on January 1, it began to look like the odds were against achieving my "golden anniversary" idea. Still, I received frequent e-mails from Shirley with enthusiastic progress reports. January 31—the actual launch anniversary—came and went. February, March, April, and the communications from Shirley became fewer, then stopped all together.

Toward the end of May, Shirley wrote to tell me she had decided to retire from her job. Immediately. Health problems had taken their toll, and she just could not handle her directing responsibilities any longer.

"As of tomorrow," she said, "Caltech will no longer have a theatre director."

Shirley said she would recommend to her successor, who had yet to be chosen, that *Rocket Girl* be produced at the school "eventually," but that she could not give me any guarantees, which I completely understood. Whenever there is a regime change in the world of performing arts, the new powers-that-be bring with them their own vision of which plays/films/art/etcetera should be promoted. The odds were extremely remote that two theatre directors in a row would be passionate about risking their reputations on a new play.

Rocket Girl, I realized, was almost certainly dead. While I waited for news from TACIT, I busied myself with other writing projects. There were dozens of them, which, like a panting Labrador, waited patiently each day for their master's attention. One morning, as I was working on a musical play composition, it dawned on me that I had never heard back from MIT.

When I wasn't writing, I would do more research, continuing to delve into the mystery of my mother's life. I use the word *mystery* because that is what it is.

Was Mary Sherman a secret agent? Perhaps she was an operative with the OSS. Or the CIA. I begin to wonder because the more I look into her life, the more evidence I find that seems to point to some sort of premeditated, intentional erasure of the historical record. It's as if Mary, or someone, has methodically and carefully expunged every record of her existence prior to her marriage to my father, G. Richard Morgan. I start asking myself, whom did my father really marry on July 29, 1951?

It gets more Hitchcockian by the day.

One day, I pick up the phone and call the records department at Minot State University (formerly Minot State College) in North Dakota. A year prior, I had had a phone conversation with the same department that said they did indeed have a record of my mother going to school at Minot in the early 40s. I decide it would be prudent, from a research standpoint, to have some sort of written documentation of her attendance, so I call and request a copy of her transcript. After running a records search they claim to have found nothing. They promise to keep searching and get back to me.

Two days later, they call back: no transcript and no records of a Mary Sherman attending Minot State College. Ever.

What happened between last year and this year?

I could have gotten transcripts from the second college she attended, DeSales College in Toledo, Ohio. Except they closed down more than half a century ago, and no one knows where the archived records are. No one knows if such records even exist.

Truth be told, many of these problems are of my own making. From childhood into adulthood I have consistently failed to understand the historical significance of certain records or events until it is too late. Photographs are a prime example. Until I began this writing project, my understanding of the importance of family photographs was meager. The house is burning—what do you grab first? Most people say, "The family photographs." Me, I just say, "Get the hell out of the house!" But telling my mother's story has changed all that. Now I kick myself for missing opportunities to preserve a handful of written and photographic records. A few years ago, my father showed me a photograph of my mother's Ray High School graduation class. It was a black-and-white 8 × 10, and it was in surprisingly good condition. As the class valedictorian, Mary was seated in a place of honor, front and center in the midst of the class. That group consisted of about sixty smiling students. Where is that photograph today? No one knows. I did nothing to preserve it, and now it's probably gone forever. At the time my father showed it to me I failed to understand its significance or historical importance. I said something like, "Oh, that's nice," then moved on to some other responsibility hounding me for attention.

There is one important score: going through boxes of old records, my father finds an 8 × 10 high-school graduation portrait photo of his future wife. Somehow, in the course of all my many writing projects, I ended up with it. It's the only known photograph of Mary prior to her wedding day in July 29, 1951. The quality is decent, but it just makes me wish somewhere along the line the Shermans and Morgans had been more diligent in taking, storing, and preserving early family photos. The urgency seems so obvious now; what was I thinking? Taking no chances, my daughter Carley and I begin a project to accumulate and digitally store all family photographs.

The house is on fire? Don't worry: everything's in the iCloud.

My family's failures to keep and maintain a proper photo collection over the past eighty years certainly explains part of the reason there is a dearth of early photographs of Mary Sherman. But eventually I become convinced that something else is at work. My mind begins to wander and I start to consider more sinister possibilities. Only one surviving photo for the first twenty-six years of a person's life? The whole thing has a mysterious secret-identity, person-without-a-name, Jason Bourne quality to it. Given that she worked in a US government top secret weapons plant during World War II, it's not so far-fetched. What kind of family, after all, keeps no photos of babies, children, and young adults? There is the high-school graduation photo, but no one avoids that portraiture duty anyway. No photographs prior to high-school graduation, two photographs at graduation, then no photos of Mary Sherman for the next eleven years—until her wedding day.

I send an e-mail to my sisters, Karen and Monica, complaining about the lack of photographic records from the Shermans. Independent of each other, they both e-mail me back and relate a story. Both versions of the story are very similar, so I give it a great deal of credence. It is a story they say our mother shared with them (but not me) many years before. We will call it the "Mother Does Not Abide Photography" story.

●─\/\/\/─●

The sun—a dependable traveler on a lonely billion-year journey—had been down for three hours. From one horizon to the other, the evening fireworks show was in full bloom. Mary lay on her back staring at it. Like oxygen slamming into the intake of a jet engine, the nippy night air flowed cold into her nostrils, warmed as it traveled down her throat, then expanded like a bellows once it reached her lungs.

As she watched the sky, ten-year-old Mary imagined a large condor lazily gliding above her. It was just a crow, but she imagined it to be a mighty condor. The previous week her teacher had taught the class a lesson about large North American birds. Mary was especially fascinated by the California condor—the largest bird in North America.

"Is it only found in California?" she had asked.

"No," her teacher had replied. "Some condors can also be found in Mexico."

"Mexico is too far. I'll have to go to California."

Mary could not tell her parents that one of her new goals was to someday see a California condor with her own eyes. She knew they would not care. Her parents would loudly disparage her, which would result in her older brothers finding out, and soon it would be just one more thing to be incessantly teased about. Mary had discovered that keeping one's dreams a secret was often preferable.

She tried counting stars for a few minutes, then gave up. There were just too many of them. Lying on her back, she whispered heavenward.

"I want to fly into space."

Of course, she knew that space travel was one goal that *was* grandiose. In 1931, humans traveling through space was solely the realm of science fiction. But it was a goal nonetheless—a goal that had set her on a new monthly routine. Several nights each month, the moon, vanquished by the peculiarities of its orbit, disappeared from the sky and revealed to humble humanity the depth of the infinite. On such moonless nights the sky would develop a display of brilliant magnificence. Each and every planet and star and galaxy would be highlighted by that perfect blackness, making their image clear and precise and glorious in their respective fixed positions. Her teacher said this lunar phase, during which the moon simply disappeared from the sky, was called the "new moon." Mary thought that was odd.

"Shouldn't it be called 'no moon'?" she had argued. "Why call it 'new moon' when there is no moon?"

"Everyone calls it 'new moon.' I can't explain it."

Keeping her eyes on the Milky Way, Mary turned her head ever so slightly to one side, allowing her cheek to caress the cool prairie grass. The fescue was over a foot tall on this part of the farm, making for excellent hiding. She guessed the time was past 9:00—putting her way overdue for bedtime. She imagined Elaine arguing with her parents.

"Why do I have to go to bed? Mary's not in bed. Why should I have to go when she doesn't have to?"

This would precipitate enough frustration for her mother to step out of the house and call from the stoop.

Mary? Mary—where are you?

But the farm was too big for her mother to conduct a search, so she would send out the boys. And if the boys found her, they would be very angry that they had to be troubled with searching for their little sister. There would be retribution, of course. Vernon might cut her hair off at night when she was asleep. Clarence might sell her next pail of milk and claim Mary had spilled it. Michael might try to shoo her horse and get it to run away.

Or they might just take a switch to her calves.

Mary knew she ought to be concerned about the consequences of rule breaking, and the fraternal retribution that so often followed, but the vastness of the stars and the eternal space beyond them were so large, so massive, so watchable. How could a person not be mesmerized by the spectacular moonless Milky Way opera as it danced and sang across the unbounded dark, a work of nature so immense and mind-numbing she could not draw her eyes away even for a moment? Was getting to bed by 8:00 really all that important when compared to watching a universe with a billion billion suns? Why couldn't everyone else be as impressed by it as she?

Then Mary heard it; a far-away forlorn call that coasted softly across the tops of the grass, moving wavelike in its harmonious assonance. The sound-absorbing quality of the prairie muffled the sound slightly, but it still managed to penetrate. Mary held her breath, less she miss it. Yes—there it was again.

Mother was calling.

"Maarryy!"

Mary waited still and quiet, hiding from responsibility as much as from a family she suspected no longer loved her—if they ever did. She listened some more, but all was quiet. Realizing her mother had given up, Mary exhaled and resumed breathing.

Then another sound—this time of movement. It was subtle enough to include nothing more than the mere rustle of half a dozen strands of grass, but that was enough. Mary looked to one side and saw Missy arrive to nuzzle her bare leg. The fur tickled against her white skin, and she smiled. The feline stepped over the right leg and nuzzled Mary's left calf—then returned to the right calf, as if unable to make up its mind which leg was more preferable. Within the quiet of the new-moon North Dakota night, the cat's purr was as clear and distinct as her father's 1932 John Deere on a windless harvest morning.

"I know you're missing your babies," she said, scratching the burr-covered feline's ears with her fingers. "You're probably wondering what happened to them, aren't you?"

With each stroke of her nails, the cat's purring increased in volume and enthusiasm. When Mary pulled her hand away, the cat pushed its head into her palm, not wanting the moment to pass without exploiting every potential caress.

"Mother called us. What do you think we should do?"

When she felt several quick, pounding footsteps thundering through the ground, she knew the boys had been released. Leaning in nose to nose with Missy, Mary held an index finger to her mouth.

"Sh."

Mary awoke.

She was chilly, and the cold had acted like an alarm clock. Still lying on her back, the first thing she noticed was that the clear skies of the previous evening had been replaced by a thick layer of clouds. Obviously her brothers had not found her, and so she had ended up sleeping the entire night in the field. Mary turned onto her stomach and raised her head just high enough to see the farmhouse. The truck was gone, meaning her father was probably in town on an errand. She could hear the "chick-chick" sound her mother clucked whenever she was feeding the chickens. Star was tied up at his post. The only person she could see was Elaine, sitting patiently on the front step, waiting for Mary to take her to school.

Mary stood up, brushed the loose grass from her dress, and ran toward the farmhouse.

"Mary—would you please put the chalk and erasers away?

"Yes, Mrs. Bowman."

Mary dutifully completed the errand, then stepped outside. Elaine was waiting.

Another school day had come and gone; time was moving so fast. With third grade almost behind her, Mary dreamed of the warm summer to come—a summer which seemed so far away under the canopy of blue-gray thunder clouds. From the top of the steps of Ray's schoolhouse she could see the Jurgensen farm several miles away. Its two-story farmhouse was highlighted by a few streaks of sunlight bursting through and settling to earth, surrounding the dwelling like a halo. As Mary watched, some dark, mystical force—angry that sunlight had punctured through its gloomy barrier—slowly closed ranks and choked off the light.

She turned away. Stepping to the ground, Mary grabbed Star's saddle belt and gave it a sharp tug to make sure it had not developed any slack. Elaine appeared at her side.

"I'm hungry."

"Mom will have something for us." Mary decided the belt was not tight enough and cinched it to the next loop. Star gave her a muffled nicker.

A woman approached them. Mary recognized her as the mother of one of the fourth graders, Bjoern Gudmund. Mrs. Gudmund wore a wide smile and carried in her hands a black-and-silver device about the size and shape of a Duckbill F-30 carburetor. It was a device Mary had heard about, but had never seen.

A camera.

"Hello, Mary. Congratulations."

"For what."

"Graduating from third grade, of course."

"It's still two weeks away."

"And Elaine—congratulations to you, too. I hear both of you have received very high marks in your class.

"Mary helps me."

"I'll bet she does." Mrs. Gudmund held out the camera for them to see. "Would you two young ladies like to have your picture taken?"

This offer made Mary's lungs suddenly inhale sharply, as if she had just overheard some naughty whisper on the playground. The Sherman family had neither individual nor family photographs, and she could not remember either of her parents ever proposing such a possibility. At a family gathering one weekend, a cousin from her father's side had suggested the Shermans sit for a family portrait. To this proposal Mary's mother had become angry and dismissed the suggestion with a curt, "I do not abide photography." Dorothy Sherman refused to allow any members of her family to be photographed for any reason, though she never gave an explanation for her position. It was a well-understood rule in their family: no photographs. Still, Mary quietly hoped to one day preserve her image with this marvelous invention. She was fascinated by the fact that someone could press a button and immediately replicate the image of whatever the camera was pointed at. The stern image of her mother's face appeared in her mind, and she decided to turn down Mrs. Gudmund's offer.

Before she could do so, however, Elaine nodded her head and moved forward. Ever cautious about angering their parents—especially where her little sister was concerned—Mary grabbed Elaine by the collar and pulled her back. She looked hard at Mrs. Gudmund.

"Mother does not abide photography."

"Your mother . . . ?" Mrs. Gudmund was uncertain what to say. "Do you mean your mother does not like to have pictures taken of her beautiful daughters?"

Mary said nothing.

"Have you ever had your picture taken? Ever? In your whole life?"

"No. Mother does not abide it."

"Your mother does not abide it," Mrs. Gudmund repeated. "Well, child. I'm not sure what to say to that."

Mrs. Gudmund considered the camera in her hands for several moments. She then turned and watched the other students as they headed off down the highway or across open fields toward homes far away.

"Mary, all I can tell you is that there is no harm in having one's picture taken. It doesn't hurt, and for your whole life you will have a record of what you looked like at this moment in time. A photograph is a whimsical token and remembrance, a historical record, a recollection of something that can never be revisited: the past. You will never again be the age you are, or look exactly the way you look now. A photograph is magical; it allows us to jump back through time whenever we please. Someday when you're older, you'll be able to show the picture to your friends, to your children, to your grandchildren. You will say, 'This is what my sister and I looked like in the year of our Lord 1930. We were sisters, we were friends, and this is where we went to school when we were your age.' Then people will look at the photograph and smile and say, 'My goodness, you were so cute, so beautiful.' Wouldn't you like to have something like that?"

Mary's expressionless face loosened up, and she almost smiled. "Yes."

Elaine looked up at her big sister. "What will mother say?"

"We won't tell her."

Mrs. Gudmund pointed to a place near the front door on which was painted the name of their school.

"Why don't both of you stand right there."

Mary and Elaine took their places where instructed and stood close together—their arms limp at their sides. Mrs. Gudmund stepped in front of them and readied her camera.

"What are you doing?" Mary asked.

"Not quite ready. Hold on."

Mrs. Gudmund lifted the camera, aiming it at the two young girls.

"How does it work?"

"You need to look happier," instructed Mrs. Gudmund. "Don't look so glum."

"How does the camera make the picture?" Mary felt she needed to know the secret of the device. A mystery was nothing more than a question waiting for an answer, and Mary began to feel that solving that mystery would be better than the photograph.

"Put your arms around each other and smile. And stand very still."

Mary put her arm around Elaine's shoulder. "How does the camera make the picture?" she repeated.

"I don't know, Mary. It has something to do with chemistry. Now hold very, very still. Don't even breathe."

Mary heard a soft click, and then Mrs. Gudmund looked up from the camera and smiled.

"That's it. When I get the picture developed I'll make sure you get a copy."

"Thank you," said Elaine.

"Thank you," said Mary.

The girls stood there watching as Mrs. Gudmund got into a pale yellow Ford and drove away. Then they mounted Star and headed back to the Sherman farm where a summer's load of chores awaited.

"What is chemistry?"

"Well that's an interesting question," said Mrs. Bowman. "What brought that on?"

"Mrs. Gudmund said cameras take pictures using something called chemistry. What is it?"

Her teacher was about to answer, then thought better of it. Instead she would give her inquisitive student a challenge. She took a small piece of paper and wrote on it the word *chemistry*, then handed it to Mary.

"You know where the dictionary is, young lady. You'll be a fourth grader soon; look it up."

Mary hefted the bulky volume to her desk and opened it to the *C* section. In a few moments she found the word.

"Read it out loud," said her teacher.

"Chemistry. The science of the composition and properties of elements and compounds." Mary looked up, as if expecting a clarification. "I don't understand."

Her teacher removed her glasses and considered the best reply.

"Your father has a truck—correct?"

Mary nodded.

"Every time your father drives his truck he is using chemistry. The gasoline in the tank mixes in the carburetor with air pulled in from outside the engine. A spark plug then ignites the mixture. That whole procedure of mixing two compounds, and then burning them, is a small part of chemistry. Understand?"

Mary nodded. "What's a spark plug?"

The teacher smiled, returned her glasses to her nose.

"I'm going to get a couple of items from the pantry. While I'm doing that, you can look up the meaning of *spark plug*."

After Mary returned from the bookshelf, she noticed Mrs. Bowman had a jar of honey, a bottle of vegetable oil, and a tall glass half filled with water. She called all the students to stand around her desk.

"Watch what happens when I pour some honey into the glass of water."

Mary, Elaine, and the other students were surprised to see the honey collect at the bottom of the glass rather than mix with the water. There were some "oohs" and "ahs." One boy asked why the honey all went to the bottom.

"Just wait," said the teacher. "Now watch this."

She poured some of the vegetable oil into the glass. Once again the students were amazed; every drop of the oil floated on top of the water. There were now three distinct layers of liquids.

"Why does it do that?" asked one of the students.

"Mary asked me earlier what chemistry is. Much of the subject of chemistry can be boiled down to this simple experiment of buoyancy and density."

On the last day of school, Mrs. Gudmund was waiting for Mary and Elaine outside the schoolhouse, a wide smile on her face.

"I have it for you."

"Have what?"

"The photograph!"

From her purse Mrs. Gudmund brought an envelope. It was not sealed.

"Straight from Kodak," she said, pulling the photograph from the envelope and handing it to Mary.

"This is what I look like? Are you sure this is me?"

"Of course, child. Why would you ask such a thing?"

"I've never seen myself before."

"Well, you've certainly seen yourself in a mirror—right?"

Mary shook her head. "We don't have a mirror."

Mary held the photo out, but Mrs. Gudmund held up her hands, palms forward.

"No, no—that's for you, Mary. You keep it."

"You're giving this to me?" She was surprised. Except for the state of North Dakota's equine gift two years before, and Aunt Aida's long-ago gift of a corn-cob doll, no one had ever given Mary anything.

"Of course. It's a present. From me to you."

"Mother does not abide presents."

Mary climbed onto her horse, and then pulled her sister up. The saddle was big enough for both of them to ride tandem, with Elaine's arms wrapped tightly around Mary's body for security. Mary guided Star away from the schoolhouse and headed down the road toward home. Star's shoeless hoofs quietly clomped along the muddy road, a road that rarely seemed to dry out. Mary held the saddle horn with one hand and firmly grasped a McGuffey Reader with the other. Her innate intelligence had allowed her to skip second grade altogether. She had achieved that advancement by squeezing time—using every available free moment for study, including the ride to and from school.

Mary no longer had to guide Star—he knew the route. From the farm each morning, down the rutted road, past the Dunkirk farm, across the river, up another rutted road, past the Swenson farm, and then a mile to the schoolhouse. In the afternoon, back the same way.

After nearly a year of this daily routine, Star was so in tune with the route that Mary could spend much of the time reading. Sometimes on the ride she would read quietly to herself, other times she would read out loud so Elaine could hear. As a result of her passion for books, Mary had become the school's best reader—surpassing students five or six years her senior.

Today she was reading a story about a boy searching for his lost puppy when suddenly the sky opened up and the sun bathed them in warmth. Mary looked up and noticed they were approaching the river, and she braced herself for the stop. Every time they arrived at the river's edge, Star came to a halt. This was the one and only place on their journey she had to take over and be the master.

"Why do you always stop here?" she said, allowing her frustration to flow from the tone of her voice.

Mary combined a gentle nudge with her heels with some verbal encouragement. That was all he needed, and Star eased across the shallow creek. As they approached the far bank, Mary noticed the river's current was unusually mild this afternoon. Once they reached the far side, Mary resumed reading. She squinted as she read, and not just due to her nearsightedness. The newly revealed afternoon sun was reflecting off the white pages, stinging and dilating her eyes. A few moments passed, then a shadow fell across the pages, and the brightness relaxed. She lifted her head and gazed skyward.

Large, thick clouds were passing overhead, moving west to east. These were not rain clouds, however, having more cotton candy than clay. Mary knew what rain clouds looked like, and these were not them. Her teacher had told her how many of North Dakota's rainstorms were the remnants of mighty Pacific Ocean tempests that had their beginnings off the coast of Alaska. Sometimes these massive storms would hurl themselves into the Dakotas with a vengeance, other times they would peter out—taking the form of giant puffy fists, separated by large, sunny gaps. The clouds today were those kinds of clouds.

"What you see there, boy, is what's left of some bigger Alaskan storm. It got played out before it reached us, and that's all that's left."

Elaine said nothing. She had gotten used to her sister's habit of talking more to her horse than to people.

Arriving at the Dunkirk farm, the girls could hear Mr. and Mrs. Dunkirk in the throes of one of their famous arguments. Sometimes those arguments went on for days. When they were finally over, it was common for neither husband nor wife to remember the original issue that had gotten the argument started in the first place.

Elaine patted her sister gently on the shoulder for attention.

"What if mother finds out about the photograph?"

"Are you going to tell her?'

"No."

"Then how would she find out?"

"Mother always finds out about everything."

"No she doesn't."

"Why do you say that?"

A puddle of water stretched across the road. The mud around it looked deep and viscous—an easy trap for Star to get a hoof stuck in. Mary steered her horse around it.

"If mom found out about everything, the boys would get into a lot more trouble."

Elaine sat quietly in the saddle for a few moments, pondering her sister's words.

"Mom and Dad don't care about what the boys do. They're all on the same side."

All was serene for a while—the cool breeze being the only sound. Mary tried to remember the last time Clarence, Vernon, or Michael had been punished for anything. She could not remember a single incidence. On the other hand, it seemed that she and Elaine were punished daily for the smallest infraction. It had been this way for so long she could not remember any other way of life.

"Why is that?" Mary asked.

"They need the boys. They don't need us."

"Why?"

"The boys do all the work."

"We do work."

"We do the work the boys don't like to do."

At that moment the Sherman farmhouse came into view a quarter mile away. A few moments of peace, then the rattling noise of a Chrysler in-line six, combined with the prattle of its flatbed's wooden stakes, rose to attention. The Sherman truck was in the area and approaching. While Mary scanned ahead, Elaine looked behind them.

"It's coming," said Elaine. "Behind us."

Mary turned to look, but her poor vision betrayed her. She needed glasses desperately, but her parents had refused to spend the money. Neither Michael nor Dorothy Sherman had ever needed eyewear, why should their children? Mary's teacher had sent them several missives expressing the problem their daughter had with her eyesight, but they would not be moved. As long as Mary could see well enough to milk a cow or clean a creamer, that was good enough.

"Who's driving?"

"Too far to tell."

"Hyah!" Mary gave Star several stiff kicks, and the horse picked up the pace, albeit more trot than gallop.

"It's the boys."

"Hyah!"

Star got the message, and a second later his legs stretched into a gallop.

"Mary!" Elaine held on tight. "Mary! We're going too fast!"

Mary took a quick glance behind them. The truck was about two hundred feet away, and closing the gap.

"Mary—I don't want to fall!"

"Just hold on!"

They could hear the boys now, taunting, whooping, and hollering. Scaring the girls was their favorite game, and the game was on. The truck's horn began blaring with an insistent repetition.

"Clarence is driving!"

Mary looked for an escape, but there was none. At this section of the road, steep irrigation ditches lined both sides—the only way out was forward.

"What are they going to do!?"

Then the truck was immediately upon them, threatening to ram Mary's horse. At the wheel, Clarence continued to sound the truck's horn, and the whooping and hollering got louder. Elaine turned her head and could see Michael sitting next to Clarence, with Vernon standing in the back, his hands gripping the stakes.

"They've been drinking," she said, holding her mouth next to Mary's ear.

The truck accelerated and pulled beside them.

"Ya wanna race!?" shouted Clarence. "Come on! Let's race to the house!"

Mary kept her gaze straight ahead—the property-line gate was not far ahead.

"Push 'em! Push 'em over!" Vernon began banging on the roof of the truck. "Push 'em into the ditch!"

Clarence turned the steering wheel slightly, bringing the truck within inches of the stirrups. Despite the danger, all three of her brothers were laughing. Mary and Elaine were terrified.

"Do it! Push 'em in!"

"Race us or you're goin' in," said Michael.

The truck nudged the stirrups, and Elaine screamed. Mary could tell Clarence was serious. Bluffing was not something he was prone to. The gate was less than a hundred feet away, but Mary could tell there would not be enough room for both horse and truck to pass through together. Someone had to yield. Clarence looked ahead, realized what was about to happen, then shouted at Mary.

"Where do you go every night!?"

But Mary kept the reigns loose.

"Where do you go, Mary!? We're tired of searching the farm for you."

"Tell us, or you're going into the ditch!" added Michael.

Side-by-side, the flatbed truck and the horse arrived at the gate. Mary prayed, Elaine screamed, and the boys laughed. Then Star— the over-the-hill feedbag that no one thought was of much value—did something he had never done before: he leaped into the air and sailed clean over the gate.

Mary pulled up on the reigns and brought her horse to a stop directly in front of the house. There was going to be trouble. She had bested her older brothers—more than bested, and that wasn't going to sit well. By refusing to yield, she had denied them their puerile victory and bruised their pride. Not just one of them, but all three of them. Simultaneously. Elaine was safe, but Mary was sure she was going to get the switch.

Mary jumped to the ground, then pulled Elaine off the saddle. She carried her younger sister up the steps and into the house, praying that one of her parents was home. She also prayed for a miracle—that today would be different. That her mother and father would break from tradition and not automatically side with the boys.

Mary set Elaine down and latched the door.

"Mom?"

No answer.

"Dad?"

Still no answer.

Through a window Mary could see Clarence was already looking for a switch, and one of the boys was pounding on the door.

By the time the girls realized their parents were not home, Clarence had entered through the kitchen door, a long maple tree branch in his right hand. Mary unlatched the front door and the girls ran outside— where Michael and Vernon grabbed them.

"You deserve this," said Clarence, and began swatting Mary's bare legs over and over. As Mary screamed and twisted, the photograph fell from her dress pocket. Clarence picked it up.

"What the hell is this?"

Mary was crying, so Elaine answered. "Mrs. Gudmund gave it to us."

Clarence roughly held it against Mary's face.

"You know how mother feels about pictures." Clarence took a closer look at the photo. "Ya know somethin'—you look ugly."

Michael leaned over to see. "Real ugly. Hey Vernon, wanna seem some real ugly girls?"

"Give it back, Clarence!" Mary reached for it, but Clarence held it just out of her grasp, laughing. Vernon stepped over to look.

"Oh yeah, they'z ugly."

"Give it back!"

Both girls were crying now.

"Please give it back."

"Oh boy—they're so ugly."

"Give it back!"

The more they pleaded, the more the boys laughed and played keep-away with the photo. Amy came around the corner of the barn to see what the noise was about. Mary turned to her older sister, tears now covering her cheeks.

"Amy! Help us! Mrs. Gudmund took a photograph of me and Elaine, and they won't give it back."

Amy shook her head. "You know Momma doesn't abide photography." Then she disappeared into the house.

"Ya want your stupid photograph back? Here!" Clarence stooped down and stuck the photo into a large wet clog of horse dung. The three boys walked away, still laughing.

"Bye, ugly girls!" shouted Vernon as they rounded the edge of the barn.

Mary picked the photo out of the gooey manure. With one hand she wiped the tears from her cheeks, then used that now-moist hand to clean the oily green slime from the photograph.

"We are ugly," Mary whispered.

Still crying, Elaine shook her head. "No we're not. We're not ugly."

"We're ugly."

"No we're not!"

Mary held the left side of the photo tight, then pulled forward with her right, ripping the photo in two.

"Mary! What are you doing!?"

"We're ugly!" she shouted.

Mary continued to rip and tear the photograph in ever smaller segments, until they could not be torn any further. Then she opened her hands and allowed the shreds to fall toward the ground. At that very instant, a cold Canadian wind rushed through the farm, lifting the photo remnants toward a gray, god-centered sky. Refusing to yield to gravity, they fluttered and flurried and scattered ever higher.

Elaine threw her arms around her sister, the Canadian wind blowing their hair over their faces.

"We're not ugly."

Their cheeks touched, and for a few seconds the chemicals of their tears mixed and slurried into a powerful new compound. Their embrace tightened, neither wanting to let go of the other. Time passed, and they eventually released. Mary turned and walked away. Elaine watched her go, watched as Mary entered the dry, brown field of last season's wheat and vanished into its crowded boscage. Though Elaine could no longer see her sister, she could hear Mary's sobbing.

Elaine called out.

"We're not ugly!"

7.

THE GREAT ESCAPE

*"Those who dream by day are cognizant of many
things that escape those who dream only at night."*
—EDGAR ALLAN POE, *"ELEONORA"*

Walter Dornberger was a mechanical engineer and captain in the German army during the 1930s. And he was one of the first of von Braun's countrymen to recognize the young man's genius and potential.[1] The moment the two men first met at a small solid rocket test conducted in 1932 near Berlin, Dornberger had decided to take von Braun under his wing. Without the captain's early help and mentorship, it is very possible history would have been written quite differently. By 1937 the two men were working together at Peenemünde, Dornberger in charge of the business end of rocket affairs, and von Braun heading up the engineering end.[2] The two men would take a little-noticed backwater technology program and build it into the pride of Germany and the technological envy of the world.

On the morning of March 23, 1939, just one day shy of von Braun's twenty-seventh birthday, Dornberger, von Braun, and a staff of over two hundred were waiting outside the offices at the Kummersdorf, a top secret facility twenty-five kilometers south of Berlin where all of the army's rocket engines were tested.[3] Kummersdorf was about to receive a visit from a very important person. Dornberger had arranged for a VIP informational tour of the facility, and von Braun had prepared a live test of two liquid rocket motors—an A-2, and its much more powerful cousin, the A-5.[4] The engineers and technicians responsible for the tests had gone over every detail beforehand to make sure there were no glitches, delays, or (god help them) explosions.

No expense had been spared for this special visit. Extra staff had been added, and the finest cooks and servers that could be found were gathered and put through a rigorous rehearsal. Dornberger had personally supervised the lunch menu and approved every item. The culinary demands of their visitor were humble—steamed vegetables and mineral water—but the same could not be said of his gluttonous entourage. After the tour of the facilities and the demonstration test-firings, their guests would dine on tafelspitz, schweinebraten, schwarzwaelder kirschtorte, and homemade apple strudel with ice cream. No detail was too small, and Dornberger personally fretted over every one.

"Remember: Say nothing about spaceflight," Dornberger instructed von Braun. "Absolutely nothing. I know how you get—all crazy and overly enthusiastic. He'll want none of it."[5]

Wernher nodded. This was only the tenth time Dornberger had given him the same tired speech. Did the captain really think he was not getting the message? Then, as Wernher's eyes meandered over to one of the new secretarial hires, he felt a nudge in his side. Dornberger made a motion with his head off to the left, and Wernher followed his gaze. He focused his eyes and saw a small fleet of vehicles heading swiftly toward them.

The Führer had arrived.

—∿∿—

Like millions before him, the man had been given no trial. Some prior prisoner somewhere had been relentlessly tortured until, in hope of saving his own life, he pointed the finger of suspicion at some other innocent Soviet citizen. That citizen did the same, and so forth, until one day a bony, tormented finger pointed in the direction of Sergei Korolev.[6] He was arrested in front of his family and taken to a secret location where he was questioned, tortured, and convicted—all in less than thirty minutes. He was then sentenced to ten years hard labor in the Siberian gulags.[7] That one of the country's most valuable and dedicated rocket designers could be so reprehensibly treated was testament to the demented and frenetic nature of Joseph Stalin's communist government.

Korolev was sent to the worst of the worst: Kolyma,[8] where it is said winter lasts twelve months and the rest of the year is summer.[9] At Kolyma he experienced a hellish world of forced labor, crowded housing, inadequate nutrition, violent confrontations, and brutal discipline. Each year, up to 10 percent of the prisoners died of malnutrition, starvation, disease, untreated health problems, murder, and execution.[10]

All of this served the purpose of supplying a steady flow of gold ore for the growing industrial might of the Soviet Union. On the day Wernher von Braun was enjoying his dessert of apple strudel and ice cream, Sergei Korolev was seated on an icy cold wooden floor in Siberia trying to keep a hundred other prisoners from stealing his only meal of the day: a small bowl of watery soup.

By the time May 1940 arrived, the death rate at Kolyma Prison had doubled, Wernher von Braun's rocket budget had tripled, and Mary Sherman Morgan was preparing to graduate from high school.[11]

On May 31, Mary gave a speech before the faculty and graduating student body of Ray High School. She earned that privilege by being the 1940 class valedictorian.[12] At nineteen years of age she was also the oldest graduate, an echo of her parents having enrolled her three years late. After the graduation ceremony, Mary and the Sherman family returned to the farm for a celebratory meal. As she sat at the table in her usual place between Elaine and Amy, Mary's mind was focused on a plan she had been formulating for months, a plan to run away from home in the middle of the night and go to college. She had told no one of her intentions, not even her one confidant, Elaine. Mary had learned long ago the advantages of secrecy, and this was one secret she had kept well. The risks of even the slightest whisper getting out were too great. She had been using the address of the local Catholic parish for all correspondence between herself and her chosen college. It was during confession one Saturday morning that she had asked the priest for permission to use the rectory as a mail drop. Since the request was spoken during confession, she knew the priest had an oath to keep the mail deliveries secret from her family.

Mary helped herself to a second portion of ham.

"My, my," said her mother. "You sure have an appetite this evening."

The year Mary graduated from high school, the farm town of Ray, North Dakota, was so small it barely warranted a stop on the one and only bus route that serviced lonesome Highway 2. Every Monday and Thursday at 3:05 a.m. American Flyer #273 pulled up to a roadside eatery known as Rachel's Diner In a typical evening, the bus would pick up no one, drop no one off, then proceed east on one of the loneliest roads in America. After Ray, it would make stops at Kenmare, Minot, Rugby, and the Sioux Indian reservation at Devils Lake. If Mary failed to catch the Monday-morning pickup, she would have to return home and spend three more days on the farm. But if everything went according to plan, she would be sleeping in a bed at Aunt Ida LaJoie's home in a couple of days. She had never met Aunt Ida, having corresponded with her only by mail. But from what other family members had told her, there was not a great deal of affection between the sisters, Ida and Mary's mother. Ida had readily agreed to keep the arrangement a secret.

The Sherman family did not own a suitcase, owing to the fact none of them ever traveled farther than fifty miles from home. For Mary, this certainly simplified obedience to the college dorm's requirement. Since her parents never saw the need to buy her anything, her clothes and personal items were few. Still, she needed something in which to carry her paltry pantry of stuff. Mary looked around for anything that could hold something.

There was nothing.

Mary slipped on her shoes, quietly tip-toed out of the closet-size room she shared with Elaine and Amy, walked through the kitchen, and stepped out the front door—holding her breath as the door's hinges squeaked just enough to be dangerous.

Mary stopped and turned around. The sound of her father's snoring was clear and steady, signaling safety. She propped the front door open and stepped onto the porch.

In the half-moon light, Mary could see Star standing quietly at his usual place, tied up at the property gate. Her father had never taken the time or expense to build a proper pen, so Mary had to have him tied up most of the time. By the time Mary had received the horse from the social worker, Star was already thirteen years old. He did not have many more good years left.

Mary emptied the last of the feedbag into Star's trough. She was about to dispose of the bag when it occurred to her that the large, durable, burlap sack would make a passable, if not elegant, carryall for her trip. She took it into her room and began stuffing it with her clothes and a few belongings. It was the early morning of June 1, 1940, and Mary was running away from home to go to college.

Rachel's Diner began life as a large trailer that had been shipped into town twenty years previous by Buddy Farnsworth. He had leased from the county a small plot of land beside Highway 2, then planted the trailer atop it. He added a wood skirt around the wheels, installed a set of glass double doors on one side and a couple of white awnings above the windows for style. He then boasted his new restaurant's original humble identity as a trailer was well hidden, a declaration no one took seriously. Buddy knew that every town—even little ones like Ray—needed places for people to gather, share gossip, and experience a good slice of apple pie. With nothing else to do in Ray, Buddy calculated his new restaurant would be a success—and he was right. He named the restaurant after his wife, who died of consumption less than a year later.

Rachel's passing did nothing to slow the traffic of hungry travelers and gossipy neighbors, and Buddy salved his loneliness with long hours in front of the stove, flipping blueberry pancakes in the morning, hamburgers in the afternoon, T-bones in the evening. In his wife's memory, Buddy kept an outdoor forty-watt bulb burning twenty-four hours a day.

It was that single lonely orb of memorial light that would guide Mary on her last mile, and she stood high on the stirrups to look for it, craning her neck for a view. But for now there was only darkness, thick and moist. The smell of the saddle and other leather riding accoutrements mingled in her nostrils with the misty summer air as Mary guided Star along the right side of the road. The only bar in town closed at ten, and in a farming community like Ray, no one ever stayed out later than when the last bar closed. At this time of night, life was so quiet and dead one could almost take a nap in the middle of the highway without fear of being disturbed.

Clip. Clop. Clip. Clop.

Sound works in mysterious ways when it has no competition. The late-night world of northwestern North Dakota was so quiet that the drumbeat of Star's horseshoes on asphalt resonated for hundreds of yards. And every insect or bird that flew, every animal that walked, made their presence known.

Mary gave Star a few soft pats on his long, coffee-colored neck. "Almost there, boy. Almost there."

Coming over a small rise, Mary could see in the distance the small number of scattered homes and buildings that made up the town of Ray. She could not remember seeing the place so quiet, so dark, so desolate. As Rachel's forty-watt light came into view, Mary took an instinctive look over her shoulder, half expecting to see her father chasing after. She gave Star a gentle kick with the stirrups.

"Let's pick it up, boy."

Mary held the saddle horn with her right hand and squeezed it tight, anxious for the events of this night to somehow move faster. Her burlap "suitcase" was tied to the saddle horn and swung back and forth in rhythm to the horseshoes. None of her brothers or sisters had felt any urge to attend college. With this surprise dead-of-night escape, Mary would be forever remembered as the black sheep of the Sherman family. And the means of that escape would be the same method of transportation that had allowed her to cross the river and begin her education eleven years before.

Mary checked her watch: 2:40 a.m. Not wanting to risk even the slightest chance of missing the bus, she had given herself plenty of time. Still, there were occasions when speed was preferable.

"Hyyeeahh!

Mary gave Star a stiff kick with the stirrups, and he obediently turned their leisurely trot into a full gallop. The remainder of the ride was exhilarating, and in short order she brought her horse to a stop outside the diner. The dull yellow bulb glowed just above a bloodred Coca-Cola machine. Mary dismounted, tying Star to the soda machine's rusty door handle. A few feet away, a pay phone was mounted on a post, around which more than two hundred cigarette butts littered the ground—a semicircle of memory marking territory for all the lonely souls who had sought solace and redemption through a telephone line.

One of the diner's two windows was crisscrossed by a lightning-strike pattern of thick tape, sealing the cracks left from a long-ago brawl. A small sign propped in the window read "OPEN" —an error that would be self-corrected in about four hours. Painted on the second window were the words, "TRY OUR HOME COOKING." Just below the window and to the left was an old wooden bench. And covering everything like a blanket was the discordant buzz-buzzing of Rachel's dying neon sign—the very first neon sign to be installed in Williams County. It was a fascination with Rachel's sign, and the high-voltage electroluminous property of neon, that had been one of the things that had piqued Mary's interest in chemistry. Buddy had purchased the sign four years previous from a traveling salesman who had promised that neon signs were not only the wave of the future, but that they would "last forever." A decade later, Buddy would discover the true shelf life of the new technology when the bright reds and purples of his sign, which had originally read RACHEL'S DINER, evolved into: ACHEL NE.

Mary pulled out a pen and paper from the burlap sack, then sat down on the bench. She began to write.

Dear Bud: Remember how your daughter Emily has been bothering you for years about wanting her own horse? I'm going away for a while. A long while. I want to give Star to Emily. She's old enough now. There are three bags of feed at my house. Tell my father I said you could have them.

Mary signed her name, then slipped the note under the restaurant's front door. She peered through the window at the glass cupboard filled with fruit pies, then returned to the bench.

Mary waited.

She checked her watch.

Time passed. The air seemed to get colder.

Ten minutes later, a truck passed.

More time passed.

Mary waited.

Then a car—a white 1945 Ford. She was surprised when it slowed, then gravel-crunched its way off the pavement and toward the diner. It certainly wasn't her father, since the only vehicle their family owned was the truck. But could he have called someone else to pick her up and take her back to the farm? Mary's heart rate speeded up, and she briefly considered running.

As the '45 Ford entered the light of the forty-watt bulb, Mary resumed breathing. The car was being driven by someone she knew and trusted—Dagmar Gudmund, a local farmer, and the husband of the woman who had taken her picture almost ten years before. In the passenger seat was his son Bjoern. He and Mary had graduated from Ray High School together the previous evening. The car stopped and both of them got out.

"Hi, Mary!" Mr. Gudmund seemed pleasantly surprised. "Wasn't expecting to see you here."

"Good morning, sir."

Bjoern removed two shiny new suitcases from the car's rear seat, then set them on the diner's first step. He nodded a hello to his former classmate as his father shook her hand.

"Where you headed?"

"College, sir."

"Really. Which one?"

Mary hesitated. Though she had graduated at the top of her class, far ahead of Bjoern, the Gudmunds had money the Shermans lacked. She knew what was coming. "DeSales College."

Mr. Gudmund hesitated just long enough to cause embarrassment. "I'm afraid I've never heard of it." That was followed by another embarrassing pause, then, "Bjoern's been accepted to NDU."

NDU—North Dakota's version of an Ivy League.

"Congratulations," said Mary, and she meant it.

"Yeah. Bjoern here's gonna be a lawyer. We're all really proud of him." Mr. Gudmund looked around, then back at Mary. "You here all by yourself?"

"Parents couldn't make it."

"Uh, huh." He turned to his son. "You want me to wait around till the bus comes?"

Bjoern smiled. "Mary will keep an eye on me."

Everyone laughed, then dad and son gave each other a warm good-bye hug.

"Love you."

"Love you, too."

Mr. Gudmund pointed to the horse, looking at Mary. "Is that yours?"

"Yes. I'm giving him to Mr. Farnsworth's daughter."

Mr. Gudmund nodded. "Well, have a safe trip." He waved good-bye to both, then got back into his car. The engine started, then Mr. Gudmund leaned out his window for a few last words to his matriculating son.

"Write often. See you at Christmas."

And with that, the Ford backed up, crunched more gravel, and headed back up the highway. After the car's lights had vanished, Bjoern took a pack of Luckies from his jacket pocket, and tapped a cigarette into his hand.

Mary was surprised. "You smoke?"

Bjoern nodded as he lit a match. "Want one?"

Mary shook her head. "No thanks."

They sat there quietly for a few moments, Bjoern smoking and Mary doing her best to stay warm. Bjoern noticed her shivering.

"You should've worn a warmer jacket."

"Don't have a warmer jacket."

Bjoern nodded as he inhaled. "The two-seventy-three'll be here soon. It'll be warmer on the bus." He held out the cigarette pack. "Sure ya don't want one?"

"I'm sure."

"Help keep ya warm while we're waiting."

Mary shook her head, and Bjoern set the cigarette pack on the restaurant step. Mary looked left and right, willing the bus to arrive. She reached into her burlap sack to check, for the hundredth time, that her money was there.

Bjoern opened one of his suitcases and removed a navy-blue jacket, handing it to Mary.

"I have two. You can have this one."

Mary shook her head. *Mother does not abide presents.*

"Take it. It's a little too small for me, anyway."

Grateful, Mary took the jacket and put it on. Her body began to warm up almost immediately.

"Here she comes."

Mary looked west, and sure enough the large headlamps of American Flyer #273 came into view. Even from this distance they could hear the gears grind as the driver downshifted twice, slowing for the stop. Mary gave the driver an enthusiastic wave, concerned that he might pass up such a little-used whistle stop as Ray. The brakes squealed, the diesel engine coughed, another gear ground its teeth, and the bus pulled into the diner's parking area. Mary removed her money from the burlap sack and held onto it far tighter than necessary as the bus came to a stop.

The bus driver stepped down from his seat, took their fares, and stuffed the money into his breast pocket. Then he loaded Bjoern's two suitcases into the bus.

"I'll hold onto mine," said Mary, clutching her sack.

Bjoern snuffed out his cigarette and climbed into the bus.

Mary stood there, her body still and unmoving. The open door of the bus was like the gaping entrance to a monstrous dark cave, full of unknowns. Everything up to this moment had been mere prologue and planning. The execution of that plan would start now—or not at all. It was not too late to change her mind—she could easily climb back into Star's saddle and return to the farm before anyone awoke, thereby avoiding the cave's unknown dangers. No one would ever know about her little midnight adventure. But the moment she stepped on that bus everything would be different, everything would change. No more cows to milk, no more creamers to clean, no more focusing on what others wanted.

No more older brothers teasing and hurting me.

There was some guilt, of course. Not guilt about leaving without telling anyone—her parents and brothers did not deserve anything better. No; the guilt was all about her little sister, Elaine. Mary would be leaving her alone, a child amongst wolves. There would be no one to stand up for her, to protect her. Once they awoke and discovered Mary had run off, they might even insist Elaine add Mary's farm chores to her own workload. They were like that.

Eventually everyone has to live their own life.

Still, it wasn't too late to turn back.

Then Mary remembered: she had written the note to Buddy and slid it under the door. Even if she returned home now, Buddy would still find the note when he opened the restaurant. Like it or not, she was committed—there would be no "take-backs," as her sister Amy liked to say.

"Are these yours?"

The question broke Mary from her reverie, and she turned to see the driver holding Bjoern's pack of cigarettes. Mary reached out and took them, placing them in a pocket of her new jacket.

"Yes. Thank you."

"Go ahead and find a seat, young lady."

"Just a moment." She trotted over to where Star was watching, his forlorn brown eyes searching for some meaning in this odd late-night escapade. It was so out of character for his all too predictable master.

"Good-bye, Star. I love you. I'll miss all those mornings you would so faithfully take me to school. You're my best friend—I'll never forget you."

Mary zipped the jacket, then boarded the bus. It was almost empty—an elderly couple about halfway down, a young black man a few seats behind them. Mary thought about taking a seat across the aisle from Bjoern, but decided against it. Had she been like most girls who were leaving home for the first time, Mary would have taken a seat at the back of the bus so she could swivel around to the large rear window, waving happy good-byes and blowing teardrop kisses to friends and family. She would cry and wave and cry some more as the bus moved on and the image of her waving loved ones retreated into the perspective horizon. But Mary was not any girl—for her the only seat was the front seat. And as she took that seat she glanced out the windows in every direction, nervous that someone or some power would suddenly appear, knife its sharp, carnivorous teeth into her legs, and drag her back to the farm.

Her legs. As Mary sat, patiently waiting for the driver to climb aboard and whisk her away, she ran her petite fingers down her legs. The physical scabs and scars of all those penitent switches were gone, but the internal scars would never go away.

The bus driver had allowed the diesel engine to idle during his stop. Now he took the two steps up to the driver's seat and took his position. The engine accelerated a few times, first gear chattered like a woodpecker, then the bus was moving. Mary could feel the vibration of the gravel underneath them, then moments later the smooth roll of the blacktop. She grabbed the chrome-plated handrail in front of her and breathed easier. She could not prevent her thoughts turning yet again to her little sister, and the feelings of abandonment Elaine would go through a few hours from now. But Mary had a life to live, and nothing was going to stop her from living it. As the driver shifted through the gearbox, American Flyer #273 picked up speed and soon transformed Ray, North Dakota, into a memory.

8.

A LITTLE OF THIS, A LITTLE OF THAT

"Three may keep a secret, if two of them are dead."
—BENJAMIN FRANKLIN

I n 1935 a group of Jesuits made a very difficult decision: to permanently close the university they had struggled for ten years to build. It was a one-building school called St. John University, and it was located just outside Toledo. Like a million other schools, organizations, and businesses, St. John had failed to survive the financial earthquake, and aftershocks, of the Great Depression. Even the voluntary work of the teaching Jesuits was not enough to save it.

A year later, a group of nuns from an order called the Sisters of Notre Dame decided to make an attempt to succeed where the priests had failed. Occupying the same building the Jesuits had vacated, the Sisters of Notre Dame founded a school.[1] They named it DeSales College, after Francis de Sales, a sixteenth-century bishop and writer. Francis de Sales was well educated, because of receiving special treatment owing to the fact he was the first born in his family. The Sisters of Notre Dame felt that de Francis's education, his scholarly writings, together with his position as cofounder of the women's group Order of the Visitation of Holy Mary, made him the perfect choice as their university namesake.

Like its Jesuit predecessor, DeSales College struggled for survival from its very first day. It had so much trouble attracting students that two-thirds of the student body was made up of novitiates, sisters, and priests. The sisters tried everything to lure a lay student body, but a one-building Catholic college with no scholastic history (and no cafeteria) was a tough sell. So when the valedictorian farm girl from North Dakota applied for admission, they granted her a partial scholarship as an enticement.

Mary sat on a step just outside the St. John Building, picking away at the small lunch she had brought with her. She wore an ankle-length plaid skirt, a white blouse, and dark-rimmed cat's-eye glasses, and her brunette locks were pinned up high in a conservative fashion. Her financial condition could be read in the torn, ragged cloth shoes she wore. In front of her ran a wide sidewalk, its entire length cracked by the seasonal freezing and expansion of ice during the legendary Midwest winters. It reminded her of the chemistry class she took in high school. A teacher had asked the class a question:

"Why does water expand when it freezes? Very few substances do that. What's so special about water?"

No one raised their hands.

The teacher called on Mary, and she took an educated guess. She knew a few things about hydrogen and oxygen atoms, and how they form special molecular arrangements.

"Water forms a crystalline structure when it freezes," she said. "This probably causes empty space, and therefore a greater size."

"That's correct." The teacher then moved on with the lesson. The other students, of course, were not impressed. Most of them had learned long ago that Mary Sherman seemed to always have the answer.

Mary looked up and down the street. It was quiet, empty. Wartime gas rationing was in effect, and people drove their cars only when necessary. The quiet was disturbed by the sound of a door on the north side of the building opening. She turned her head and saw two nuns from the Order exit the building and head up the street toward the convent.

Mary enjoyed attending DeSales; she felt comfortable studying alongside so many priests and sisters. She was beginning to rediscover the religion of her childhood, and there was no better place to be for that than DeSales College.

Mary checked her watch—ten minutes until Dr. John Thornton's chemistry class.[2] Chemistry was still her favorite subject, though it was difficult to explain why. Mary herself was unsure what the attraction was. She had an aptitude for science and math in general, but her mind just seemed to fire on all cylinders when the subject was chemistry.

Mary saw movement above her and looked up just in time to catch a hawk gliding in for a landing on one of the turrets atop the college roof. A pair of hawks had made a home there, a spaghetti mix of straw, thread, and bramble garnered from a nearby dairy farm forming their nest. Mary watched as the hawk fed a morsel to a small chick.

Why couldn't you be a California condor? Mary answered her own question out loud. "Probably because we're in Ohio."

Mary reached into her lunch sack and retrieved a small note. Father Lyons had handed it to her in the hallway that morning. A man wanted to meet with her about a job interview, and since nonstudents were not allowed in the building, she had been instructed to wait on the front steps. She looked up and down the street again—still no one.

Mary had never had a regular paying job. Sometimes she dreamed of how much money she might have if her father had had to pay her for all those years of farm work.

Over the past year and a half, her family had gotten used to the idea that Mary had a personal need and drive to go to college, even though no Sherman family member had ever seen the need or had the drive before. Of course, that did not mean they were willing to financially support her. Mary had no choice but to find some income. The money from her $500 scholarship had been exhausted, and the Sisters of Notre Dame, good Christian women every one, needed a payment. Soon. Whatever this job was—and Father Lyons had been very cryptic about it—this was one appointment she could not miss.

The north door opened again, and about a dozen students walked out of the building. Most of them were women, of course—the war having removed from American soil a large portion of the country's young male population. She watched the students cross the street, then head east toward Guy's Café, the closest place in the neighborhood where one could get a bite to eat. Mary had only taken advantage of it once—her finances allowing little else. As she watched the students head up the street, she saw a man maneuver through their midst heading the opposite direction toward the school. He was very thin, with a bony face. He carried a briefcase and was impeccably dressed in a dark suit and felt hat. The man crossed the street, checking his watch as he approached.

Mary pushed the last vestige of her sandwich into her mouth as the man came to a halt just below the step she was sitting on.

"Hi. I'm looking for one of the students here. Mary Sherman."

"That's me," she said, her mouth still masticating the bread and turkey meat, with a little mayo and lettuce.

They shook hands.

"I'm Paul Morsky. We have an appointment."

Mary nodded. "I've been waiting for you."

"Sorry I'm late. Is there a place we can talk?"

"What's wrong with right here?"

Paul glanced to his left, then to his right. Mary thought he seemed nervous.

"I'd prefer someplace with a little more privacy."

"You're not allowed inside the building." Mary considered their options, then added, "There's a public park up the block."

They didn't speak during the short walk, and when they arrived Mr. Morsky took a seat at a wooden picnic table. To Mary the man seemed nervous, taking occasional glances left and right.

"Are you nervous?" she asked.

"No. Just careful. Loose lips sink ships."

"Huh?"

"Never mind." He removed his felt hat and set it on the table. "Thank you for taking the time to meet with me."

"I have to get to my chemistry class soon."

"Chemistry. You enjoy that?"

"It's my favorite class."

The man smiled for the first time.

"Good. What I have to say will only take a few minutes," he said, handing her his business card. Mary read the company name.

"Plum Brook Ordnance Works." Ordnance. Something to do with

explosives. She remembered one of the students saying her mother and sister worked at Plum Brook.

"A lot of girls here in Toledo work for us," said Paul. We'd like you to consider coming on board."

"What would I be doing?"

Again with the look left, look right.

"What would you be doing? Let's just say, 'a little of this—a little of that.'"

Mary sensed his coyness was deliberate. "What do you do for them?"

"I'm a recruiter. My job is to find people—people with certain talents."

"How did you get my name?"

"One of your teachers referred you."

"Which one?"

"I'd rather not say."

"Why me?"

Paul Morsky shifted position on the bench, taking a moment to think through his reply.

"The war has hurt this country, Miss Sherman. More than most people realize. We have a great many skilled jobs that are not being filled."

"Like . . . ?"

"Like chemical engineers, for one. There's a big shortage of trained chemists right now."

"You're aware I'm just finishing my sophomore year here."

"Oh yes—we know all about you." He opened his briefcase and removed a thick folder. "Your teachers give you glowing reports. You're at the top of your class in both math and science."

"This isn't exactly Harvard—not a lot of competition at DeSales."

Paul chuckled quietly. "You're modest. I like that." He paged through the folder briefly, then re-closed it. "How would you like to skip college and go straight into your career? We'll hire you as a chemist—right now. Not quite the same pay scale as if you had a degree, but close."

"You must be desperate."

Whatever Morsky was thinking at that point, he did not say.

"Mr. Morsky, I'm looking for employment so I can continue my schooling, not abandon it. I didn't graduate from high school till I was nineteen; I've already had too many delays in my education. I don't want any more."

Though the job offer was a compliment, she handed him back his card. "Sorry. I'm determined to finish my degree. I'll call you after I graduate." Mary was about to stand up when Morsky reached over and held her arm. In her mind she pictured Clarence, and he had a switch.

"Please let me go."

"Listen to me, you selfish little girl. There are a million American boys out there who have postponed their educations for years in order to keep the world safe for people like you and me. A lot of those American boys are dying out there. Every day hundreds of our young men perish on the battlefield. Don't you think you owe it to them to contribute something?"

He paused to allow his words to sink in to her conscience, then he released his grip.

"Your job would help the war effort, Miss Sherman. Our soldiers need you."

"Do you use that guilt technique on everyone?"

"Only if I have to."

"Does it work?"

"Only with the good people. And I believe you're one." Mr. Morsky handed her back the business card.

Mary thought of her brother Michael, who had joined the army the year before and who was at that very moment on some battlefield somewhere risking his life for democracy. After all the years of torment they had put her through, she had no great affection for any of her brothers. Still, Mr. Morsky's words had cut her to the center. Was she being selfish? Undeniably, but only because she had never before had an opportunity to put herself first. All her waking days had been spent serving someone or something else. Was it so bad to be a little selfish after all she had endured?

"I suppose I could finish my education taking night classes."

"Sure," he said. "You could do that."

Had Mary been paying better attention, she might have detected the doubt in his voice.

"What do I have to do?"

"You'll need to apply for a top secret clearance—just a formality in your case. Aren't too many Soviet spies coming out of Ray, North Dakota, these days. Takes about a month, so you'll be able to finish out the school year. After that, you show up for work—seven thirty in the a.m."

Paul stood up and once again offered his hand. She accepted it.

"Congratulations."

He turned and took two steps toward the street.

"Hey," she interrupted. "You haven't told me what I'll be doing."

Paul looked carefully at his new recruit. His reply had an edge to it—a mood; like two spies exchanging whispered secrets in a Moscow alley.

He slowly placed his felt hat back on his head and said, "A little of this—a little of that."

Then he turned, walked across the street, and was gone.

As she headed away from the park toward a rendezvous with chemistry class, Mary held her books close and kept her head down, counting each crack in the pavement. She did not want to interrupt her education—again. She loved school more than life itself. Leaving the farm and striking out on her own had been the best decision she had ever made. While many of her fellow students at DeSales battled constant waves of homesickness, Mary rarely thought of her pre-college life. Free to make her own decisions. Free to make her way in the world. Free to make her mark. Other girls could grow up to be schoolteachers—that was fine for them. Mary had other plans.

Still, freedom had its limitations. Mary's paltry savings account had run dry, and she was desperate. It's when someone leaves home that they discover how important money really is. If you cannot buy food, you starve. If you cannot pay the rent, you become homeless. Mary was close to suffering both. To top it off, the school's bursar kept putting notes in her mailbox, emphasizing in ever stronger terms the need for Mary to bring her tuition payments up to date.

Then, out of nowhere, a job offer. A job that would allow her to start her career immediately, but whose description was couched in riddles. A job that promised to end her financial problems, but whose acceptance required a top secret government clearance.

A little of this—a little of that.

As she approached the entrance to DeSales, Mary happened to look up and see Mother Hawk attending to her chicks. Then Father Hawk, back from a foraging mission, flew up and landed, another tasty morsel in his beak.

Even birds have a better family life than I do.

It was getting late in the day, and the St. John Building was casting a long, cold shadow. Mary sped up her pace—being late for class was not something she was known for. Arriving at the north door, she took one last look around the grounds. She had enjoyed her brief sojourn at DeSales College so far, and hoped to be able to continue. Mary, after all, had never given up on anything in her life and certainly did not want to start. But change, like the persistent force of a Dakota prairie wind, was blowing fast and strong. Mary entered the building and closed the door behind her.

—∿∿—

"Miss—your booties."

Mary turned to see who was talking. It was the receptionist, a mousy woman with stringy hair and a bow-tie kerchief wrapped loosely around her neck.

"Excuse me?"

"You cannot enter the floor without booties."

"I—I don't have any."

"Are you new here?"

"First day."

The mousy lady stood and opened a faded green metal cabinet

about six feet tall. Inside were what looked like hundreds of pairs of white socks. The lady removed a single pair, closed the cabinet, and handed them to Mary. She motioned to a row of chairs where she could put them on. Above the chairs hung a number of black-and-white portrait photos of Plum Brook Ordnance's board of directors. On another wall was a poster showing a sinking battleship. Around the half-sunk ship were the words LOOSE LIPS MIGHT SINK SHIPS.

"No one is allowed on the floor without booties." She pulled a padlock and key from another drawer. "Pick the first available locker you find and place your personal effects in it."

Mary sat down. The wooden chair was stained a shade of deep brown, like the bark from a South Dakota spruce. The booties had an elastic band that allowed them to fit over her shoes. She put them both on, stood up, and faced Mousy Lady.

"Do I have to work in these?"

"Of course."

"What are they for?"

"They help dissipate static electricity—so you don't blow the place up." She put a cigarette in her mouth and kept it there, a signal she was done talking.

Mary walked to the large set of double doors labeled BADGES REQUIRED BEYOND THIS POINT. She knew her Plum Brook Ordnance ID was still clipped to her blouse, but looked down to check just the same.

"What am I going to find beyond these doors?"

Mousy Lady set her cigarette on an ash tray, then began turning the platen knob on her typewriter, feeding through it a three-color carbon form in triplicate.

"A little of this—a little of that."

The woman's reply made Mary instinctively glance at the "Loose Lips" poster. Then she turned back and twisted the knob on the door, pushing it open. Three steps later, she was standing in front of a small wooden desk manned by two burly security guards.

"Your ID, please."

Mary unclipped her badge and handed it to the man while the second guard walked around the desk to inspect her booties and check the contents of her purse. The first guard spoke to his companion.

"Tell Hollingsworth his new chemist is here."

"Right," and he was gone.

A moment later, Mary had her badge back and the guard waved her on.

"I'm supposed to get a locker."

"First door on the left."

The locker room was larger than the Sherman farmhouse, with hundreds of pale tan lockers. Two older women were talking in low tones against the far wall. They looked up when Mary came in, watched her briefly, then went back to their business. There were dozens of available lockers. Mary picked one at random, then put her purse inside. That's when she felt a hand on her shoulder.

"Miss Sherman?"

She turned quickly to find a short, middle-aged man standing before her. Mary wondered what a man was doing in the women's locker room.

"It's not that kind of locker room," he said, reading her concern. "The only things people stow in here are their personal valuables. And, of course, metal jewelry. Can't wear any metal jewelry on the floor."

Mary looked at her watch, then removed it, placing it in the locker beside the purse. Then she closed the door and secured it with the padlock. The two of them shook hands.

"A. J. Hollingsworth—Floor Supervisor. Welcome to the team." He smiled and headed for a second door. A very large sign across it read NO SMOKING—BOOTIES REQUIRED.

"Come on—let me show you the place."

The "floor" was a cavernous room larger than an aircraft hangar. Mary guessed there were at least four dozen workers whom she could see from her vantage point. There was a cacophony of hums and roars and people shouting. Giant machines—all painted a putrid shade of green—were everywhere. Several odors permeated the air: sulfur, ammonia, and something else she could not quite place. A large sign with letters more than two feet high read, ONE SPARK KILLS EVERYONE. Metal poles five feet high were placed at regular intervals. Whenever an employee passed one, they would touch it, then move on their way.

J. pointed to a nearby pole. "Please touch that."

Mary walked over and touched the pole.

"Thank you. Get in the habit. Always touch a pole when you pass one so you can discharge any static electricity buildup." He stepped farther into the room and spread his arms wide. "Welcome to Plum Brook Ordnance."

"Now that I'm here, can someone finally tell me what this place is about?"

J. turned to look at the factory with a face full of pride.

"Miss Sherman—you're looking at the world's largest manufacturer of trinitrotoluene."

She had only studied chemistry for three semesters—one in high school and two at DeSales, but it was enough. Enough to know what all the prefixes and suffixes in that word meant. Enough to know the identity and purpose of the chemical compounds. Enough to know exactly what all the mystery and fuss and secrecy over Plum Brook Ordnance had been about. She looked at Mr. Hollingsworth and whispered.

"T-N-T."

"We also make a helluva lotta gunpowder. Also nitrocellulose, nitroglycerine, some cordite. But TNT is certainly one of our most popular products. I'm sure you're familiar with the story of Alfred Nobel."

"Nineteenth-century inventor. Founded and funded the Nobel Prizes in his will. Made a fortune as the creator of several well-known explosives. No self-respecting chemistry student would be caught not knowing the story of Alfred Nobel."

"Very good. Let's take a tour." Mr. Hollingsworth stepped farther into the factory, and Mary followed. "TNT is made from a mixture of one mole dinitrotolulene . . ."

". . . three moles of concentrated nitric acid, and five moles of sulfuric acid," she finished.

Mr. Hollingsworth stopped walking. "Yes. That's quite correct. The mixture is then heated to one-hundred thirty degrees centigrade for . . ."

"Do you still use water to filter out the solution?"

Mr. Hollingsworth seemed disappointed—like a doctor whose patients had all cancelled their appointments. "Yes—that's still generally considered . . ."

"After you distill it with sodium sulfite and wash out the impurities, what do you do with the waste water?"

The supervisor shrugged. "We pour it into a big hole in the ground out back."

"Exposed to the air?" Mary was incredulous. "That hole must contain thousands of gallons of sulfuric acid."

"Millions. It's a pretty big hole."

"I can't believe the government would let you do that."

"Actually, it was the government's idea." Hollingsworth extended his arm to show the way. "Shall we continue?"

The tour took less than thirty minutes, and after it was over, all Mary could think of was the many ways she could help her new employer improve their operation.

"Let me show you where you'll be working."

He led her to a door labeled "INVENTORY TESTING." The white paint on the door was yellowing and peeling. The yellow tinge resembled a typical oxidation reaction, and Mary guessed it was the result of the paint reacting with the chemicals that permeated the factory air. As Mr. Hollingsworth searched through a large bundle of brass keys, she asked, "Why don't you have your workers wear protective masks?"

He shook his head. "Law doesn't require 'em." He found the correct key and unlocked the door. "Besides, the ceiling fans pull out most of the fumes."

As he opened the door, Mary glanced up. About a half dozen small fans were mounted in the ceiling, three of them working. The other three appeared to be rusted in place.

"Come on in."

The room had two long wooden tables with Formica countertops. There was the usual assortment of Bunsen burners, beakers, and ovens. A large blue test tube centrifuge sat in a far corner. On the wall to her left sat a metal cart. A large number "96" was painted on the wall above it. On the far wall was a similar cart, above which was painted the number "92." She looked up to check for ventilation fans. There were none, but what she did find were large swaths of paint bending away from the ceiling and curling downward, each with that same yellow hint.

"What will I be doing in here—and please don't say 'a little of this, a little of that.'"

"You'll be testing the incoming nitric-acid shipments." He pointed to a long row of twenty-five-gallon barrels lined up against the far wall. "A lot of what we make here requires nitric acid as an ingredient, and the acid purity must be at or above 96 percent. Absolutely essential."

He pointed to the carts.

"Any barrel that tests ninety-six or better gets put on that cart. Ninety-two or better goes on that cart. We can use the ninety-two for some applications."

"And if it's under ninety-two?"

Mr. Hollingsworth opened a drawer mounted beneath one of the Formica counters and pulled out a pad of large orange stickers. Each sticker read, "REJECTED FOR IMPURITIES."

"If that happens, slap one of these puppies on it and let me know— I'll have it sent back to the distributor." He spread his arms wide to indicate the entire room. "It's your baby." He turned to leave.

But it was all too much for Mary. To open, test, and analyze high-strength nitric acid in a small, enclosed, unventilated space like this; it was very hazardous—almost suicidal.

"You can't be serious! You expect me to spend eight hours a day working with highly pure nitric acid—whose fumes are extremely toxic—without any masks or safety gear? What kind of place are you running here!?"

Mr. Hollingsworth seemed hurt. He went to a metal cabinet mounted against the wall by the door and opened it. Inside were several gray jumpsuits, gloves, boots, glass air masks, and backpack oxygen bottles.

"I recommend you use some of this stuff." He opened the door to leave, then turned around. "And as far as what kind of place we're running here; Plum Brook Ordnance is a crucial part of the war effort, Miss Sherman. Everyone who works here is helping our troops liberate Europe. Is that important enough for you?" He stepped into the factory proper and shouted above the din. "We need all those barrels tested by the end of the day. Lunch is at eleven-thirty."

He closed the door and a modicum of quiet returned.

Mary noticed all the gray jumpsuits were the same size. She found the one that looked to be in the best condition and removed it from its hanger, holding the suit against her body. It became clear right away this equipment had not been designed to accommodate five-foot-five, one-hundred-eleven-pound young women. Putting this thing on would swallow her up. She set the suit down on the table, folded her arms, and began to walk the room. Surveying her new work environment, Mary rounded the second table, glanced at the acid barrels, and briefly checked the equipment. Arriving at the blue centrifuge, she detected a ripping sound behind her and turned around just in time to see a car-size sheaf of ceiling paint float gently downward and land on the far table.

I hope the war in Europe is going better than this place, she thought, giving the centrifuge a moderate spin.

Still, a job was a job was a job, and she needed it. Mary consoled herself by remembering that working in her chosen field, no matter what the conditions, was better than waitressing—the employment most of the other coeds at DeSales ended up with.

Returning to the jumpsuit, she put one leg into it, then the other, then both arms. Zipping it to her neck, she almost laughed at how the suit sagged on her tiny frame. Next were the ankle-high boots, the oxygen bottle, the facemask, and finally the gloves. It took a few moments to figure out how to get the oxygen into the mask, then she spent a minute testing the gear. With the glass mask covering her entire face, the sound of her breathing became accentuated.

Confident the equipment was working properly, Mary turned toward the wall and lumbered robot-like toward the first acid barrel. As she approached, she thought back to the first time she had worked with nitric acid in her chemistry class back at DeSales. It had been a small one-pint bottle, and she would never forget its warning label: CONTENTS DISSOLVES HUMAN FLESH.

And that was only a 30 percent solution.

The tops of each twenty-five-gallon barrel were sealed with a locking steel band. Pulling on a handle released the locking mechanism and loosened the band. She deftly pulled the band off and set it atop the second barrel.

Then, with both hands, she removed the lid and looked inside.

9.

AN ODD NUMBER

"All parents damage their children. It cannot be helped. Youth, like pristine glass, absorbs the prints of its handlers."
—MITCH ALBOM, *THE FIVE PEOPLE YOU MEET IN HEAVEN*

My parents have asked me to help them run a garage sale. When they say they want me to "help," what they really mean is they want me to do all the work. I live closer than any of my siblings, so I'm the one who usually gets these calls. Since my father has diabetes and Parkinson's disease, and my mother has heart problems and emphysema, I feel guilty if I don't readily agree. My mother has just turned eighty-two, and she has fifty feet of oxygen tubing attached to her body. It follows her wherever she goes. She stopped smoking years ago, but by then the damage had been done.

I don't mind helping my parents with these kinds of projects—that's not the problem. The real problem is that they are incapable of dealing with certain kinds of reality. Their house is filled with half a century of accumulated possessions, almost none of them worth more than a few dollars, though they don't see it that way. For example, they have a cheap, off-brand backpack left over from their Sierra hiking days. It's in horrible condition. They paid seventeen or eighteen dollars for it forty years ago, and they expect me to sell it for ten. They would never pay that much for it now, yet they expect me to sell it for that. I know all of this, of course, when I agree to help, so I go into it with eyes open. Still I marvel at how the two brilliant rocket scientists I call my parents can solve intense math and engineering problems on one hand, yet fail to understand the simple economics of a worthless backpack on the other. They're not alone; many people have this problem, including me. The reason is simple: an old, beat-up backpack is never just a backpack—it's a repository of memories of all the places it's been. Aren't those wonderful memories worth at least ten bucks?

Yes, but only to those who created the memories—the original owners.

I go to their house in Canoga Park and open the garage door. Everything they want to sell is in a pile in the center. They've put price tags on how much they want each item to go for. Not one single gimcrack or knick-knack is priced to actually move. But a promise is a promise, and I start laying items out onto the driveway. We had placed an ad on Craigslist the previous week, and now people start arriving and going through the junk. I can see their eyebrows furrow when they see the price tags. My sister Karen shows up, having driven two hours from San Clemente. She has a CD player with speakers she doesn't want anymore. She puts a price tag on it and sets it with the other items.

It's the first item to find a buyer, and sells in seconds.

As she and I laugh at how much more valuable her stuff is than our parents', I continue putting more items out for sale. I come across an old game of Trivial Pursuit—one of the original ones from 1979. It brings back a flood of memories. I had bought it for my mother for her birthday since I knew how much she enjoyed any sort of intellectual game. The first night our family opened the game and played it turned out to be the last night as well. Our mother knew so many of the answers that she dominated the match, slaughtering everyone. Nobody wanted to play with her after that experience, and so the game went into permanent closetry. My parents had piles of stuff like that: trinkets and trifles too valuable to throw away, not usable enough to keep.

The day wears on and we manage to sell a few items. By the time we're done, the cashbox has less than $30 in it, half of which came from Karen's stereo. She and I each chip in twenty bucks. While I put most of the remaining items into my SUV to throw away later, Karen tells our parents how we sold almost everything.

"See," my dad says. "I told you that stuff was worth money."

As I get in my car to go home, I retrace my route and take down all the "Garage Sale" signs I've taped up around the neighborhood. An hour later, I arrive home, boot up the computer, and start paging through the voluminous pile of family records I have accumulated. So much of what I have has yet to be read, let alone closely examined.

The public library in Sandusky was always quiet on Friday evenings. Most everyone would be off at some party or movie or restaurant. Mary looked around the room; there were a couple of high-school students doing homework. Two tables away was a skinny guy who had been rejected by the Selective Service due to some foot problem. At the end of her table was a young woman of college age reading a textbook and making notes with a number-2 pencil. Mary recognized her from many previous visits to the library.

Mary turned the page of the chemistry textbook she had brought with her from DeSales, arriving at the chapter titled "Gas Stoichiometry." She stared at the page, willing her mind to read, study, remember. She wanted to study, she needed to study. There was so much she needed to know in order to perform her job. If only she had listened better in class, studied a little harder.

Focusing on learning more chemistry, however, was blocked by a peculiar math conundrum that persisted in superseding everything chemical. Numbers kept flowing into her conscious—one number in particular, buzzing around inside her head like a giant bee, bewildering her with its weird coincidental appearance in all facets of human knowledge both scientific and historical. It kept popping up in astronomy, it kept showing up in chemistry. It made occasional appearances in religion and archeology. In math and physics, it was just plain ubiquitous.

And, of course, there was the biological connection.

How could I have been so blind not to notice this number before?

Everywhere she turned, there it was: some new application, some novel convention, some heretofore-unknown praxis of the number.

The number 28.

Through one of the upper windows Mary could see the moon framed perfectly. A waxing gibbous, she was pretty sure. Was it waxing or waning? Hard to remember. Too many distractions right now, especially the distraction of coincidence. Mary had never considered the coincidence before, but there was no escaping it. Twenty-eight days—what an odd number. An even number, to be sure, yet odd in its own way.

She was late—too late. There had been a few one-day-lates over the years. A couple of two-day-lates. Once, after a particularly grueling, hot August, she had been a whopping three days late. But never twelve. Twelve days late? Never. And what exactly was going on with her complexion? Every time she passed a mirror she would stop and look. Something seemed to have changed ever so slightly. She would lean in, squint, turn her face left and right.

What's going on with my skin?

Mary had weighed exactly 111 pounds ever since graduating from high school.[1] The scale never swayed more than a couple of ounces in either direction. She was pretty sure that was about to change. Soon she would be gaining weight.

Was it a coincidence that a woman's menstrual cycle matched almost to the minute the phases of the moon? How did that come about, exactly? Was it the result of some sort of ancient evolutionary necessity that had, over the years, become the biological body-clock equivalent of an appendix? Or was there no connection at all.

28.

Mary turned to the girl at the end of the table. Since there were only five visitors in the library, there was no urgent need to be quiet.

"Did you know that twenty-eight is the second perfect number, the first being six, of course, owing to its relation to the Mersenne prime seven. You would have to count to 496 before arriving at the third perfect number."

The girl stared at Mary. "Are you a math major or somethin'?"

"I used to be a chemistry major."

"Used to be?"

"I quit school to work at Plum Brook."

The girl nodded. "Yeah. My aunt works in the shipping department." The girl returned to her textbook, scratching off more penciled notes.

28. As everyone knew, 28 was a harmonic divisor number and could be found in the Padovan Sequence. A one-year calendar based on thirteen 28-day months would simplify annual timekeeping immensely, but so far no modern culture had adopted such a calendar. Even so, the math to support it had always been there. But most interesting of all, 28 was the only positive integer with a unique Kayles nim-value. And there was more: 28 was the number of days required for the sun's exterior to make one complete revolution around its core. What on earth was so special about the number 28?

Twenty-eight.

Rhymes with late.

Mary skipped past gas stoichiometry, flipping ahead to chemical kinetics and thermodynamic equilibriums, finally letting the book fall open at the chapter marked "Metaloids." Without reading the text, she mentally recited the metalloid elements: boron, germanium, arsenic, antimony, tellurium, and silicon.

Silicon. There was some news recently about silicon. Mary remembered reading an article in a paper or magazine about a scientist who claimed that silicon would be the "element of the future." Something about its semiconductive properties had significance.

28 . . . 28 . . . 28 . . . 28 . . .

Mary shook her head several times, as if physically rattling one's brain could dislodge unwanted thoughts. There were more important things to consider than coincidental numerology. She needed to get her mind onto something constructive, so she located the subheading for silicon and started reading.

"Why can't we ever see the other side of the moon?" Mary was in her freshman year in high school. No more one-room-schoolhouse lifestyle; Ray High School had over three hundred students. It was a much different world. She was frustrated, however, by her science teacher who frequently was unable to answer her questions.

"The reason we never see the far side of the moon is because the moon makes one complete revolution in the time it takes to make one full orbit," he said.

"Yes, yes, I know that," said Mary. "But why? Why does it make exactly one revolution about its axis in the time it takes to make one revolution around the Earth? And it's been doing that for thousands of years. Probably millions."

"What's your point, Mary?"

"My point is that it's an odd coincidence. Don't you think? If the revolution of the moon on its axis were off by just a few seconds, people on Earth would eventually be able to see the far side of the moon. But the two revolutions correspond exactly. Absolutely exactly. How can you stand that coincidence?"

The other students giggled. This was just Mary being Mary.

"We're studying geology this week. Why don't you bring this up when we get to astronomy?" The teacher then continued his lesson on the formation of quartz crystals.

Eventually Mary would learn about planetary tidal forces, and how those forces easily explain the synchronization of the Earth/Moon system. But until that day came, she would be forced to live with daily frustration. Too many questions, not enough people who could answer them.

Silicon.

Silicon was discovered in 1824. Its original name was silicium—a name retained to present day by some countries. It is the eighth most common element. Like carbon, it is mostly inert. It is a solid at room temperature.

"Its atomic weight is twenty-eight," she whispered to herself.

28 . . . 28 . . . 28 . . .

Twenty-eight.

Rhymes with weight.

Mary had stopped at the nurse's office that afternoon.

"You're up a pound," said the nurse. One hundred twelve."

Mary had never been over 111—ever. Now she was 112. 112 was a new number.

"Did you know," she asked the nurse, "that twenty-eight divides into one hundred twelve evenly—exactly four times?"

Mary heard a chair move and noticed the skinny student had stood up and was gathering his books to leave. A few moments later, Mary heard one of the large security doors in the back of the library close, echoing throughout the building. A librarian was pushing a cart full of books to a rack in the sociology section—other than that, Mary guessed she and the college girl were alone. She looked up at the clock.

8:28.

In four more hours she would be thirteen days late. Technically.

It was 1943, the middle of what one of her former classmates at DeSales liked to call the "Send-Away Generation." Her name was Kathleen McNulty, and she was always lecturing Mary about "being careful," otherwise she would become a member of the Sent-Away Club. They had laughed at it then, but no one was laughing now. It was because of Kathleen that Mary discovered bridge. They had been together one evening listening to the Richfield Oil Company Variety Hour—a radio program involving live music and celebrity guests. One night a week the station would host American bridge expert Alfred Sheinwold, who would take a few minutes discussing the intricacies and strategies of his favorite card game. Kathleen had gotten Mary to listen with her, and together, using nothing more than Sheinwold's radio show, they taught each other how to play bridge. For Mary, it would be a life-changing experience.

But that change and that life was still years away. For now, Mary had much bigger concerns than which suit was trump.

Ever since the end of World War I it had become fashionable to "send girls away" when they became pregnant, a sociological phenom-enon whereby a young, unmarried pregnant woman would be put on a bus and sent to live with a friend or a family member hundreds of miles away. This was done in order to, supposedly, prevent shame from infecting some good family name or staining some employer (or institution of higher learning). Meanwhile, back home, everyone would pretend the girl was on a trip, away at college, visiting family, or some other fools-nobody code phrase. All it really did was create more fodder for gossipy tongues. The fact that cover stories never prevented aftermath gossip was testament to the fact that gossipy tongues were, and always had been, a potent, unstoppable force, invincible to any power that might be foolish enough to try to thwart their process. Every month or so at Ray High School, and later at DeSales College, a co-ed would simply vanish. There on Tuesday, gone on Wednesday. And the students would start to whisper.

Did she have a boyfriend?

They did it in a car.

I heard she was raped.

That's not what I heard.

I always knew she was a slut.

Whispers, whispers, whispers.

Mary had never considered such a fate would ever befall her, yet here she was, having a date with twenty-eight. Now she had to seriously contemplate the possibility: would she become a card-carrying member of the Sent-Away Club?

Sent-Away? No, that's not it.

The more Mary thought about it, the more she realized girls weren't really sent away—they were hidden away. It was all about hiding things—hiding pregnancies, hiding babies, hiding adoptions, hiding people, hiding shame, hiding reputations, hiding one's face, hiding the past, hiding the future, hiding the truth. Yes, truth. The "Sent-Away Generation" was really the "Hiding-the-Truth Generation." Hiding truth, all in the name of saving one's golden reputation. All of it based on some flawed presumption that it was possible for a chalked-up slate to be utterly cleaned. Forever. Mary looked over at a chalkboard nailed to the library wall. Someone had thoroughly cleaned it with soap and water, yet chalky streaks and residue remained, forever marring the original slate that had come from the factory, clean and pure.

The pretense of all this was that the past was not the past—that one could make the past disappear by denying its existence. The past was all just some bad dream that could be easily erased by buying a bus ticket to Aunt Kathy in Kansas. If no one back home knows you were pregnant, then that means you were never pregnant. If no one knows you had a baby, then the baby doesn't exist. Ignorance is bliss.

His name was Patrick—a cute, athletic Irish boy. Mary fell for him immediately. Patrick was everything she was looking for in a boy—handsome, smart, and Catholic. They met in English Composition 102 at the end of the second semester. It started the way a billion prior human relationships had begun; he smiled at her, she smiled back.

They began dating, if that was what it could be called. Neither of them had much money, so a date would often be nothing more than strolling around the block, hand in hand. He had a sense of humor and treated her like a lady, both of which she was especially appreciative of. After a lifetime of enduring merciless teasing and endless taunts from her older brothers, after nineteen years of never hearing the words "I love you" from her parents, after wetting her pillow night after night and praying to God to deliver her the smallest morsel of human kindness, Patrick arrived on cue. So desperate was she for a gentle word, a soft touch, a warm embrace, that her otherwise-analytical mind shut down and she surrendered to him whole. She fell hard, like a dry-docked ship's anchor hitting cement. He pulled her into his vacuum, and she offered no resistance.

Even after Mary had dropped out of DeSales and started working for Plum Brook, they had continued their relationship, as best they could. Sandusky was only thirty-five miles from the college, but due to wartime gas rationing, their opportunities to be together were limited.

One day Patrick said, "I love you," and before long Mary noticed her facial complexion was changing. Nausea became frequent, and each day began with a ceremonial vomiting of her breakfast coffee.

Owing to the dangerous chemicals being used and manufactured, Plum Brook Ordnance had an infirmary on the premises. One day Mary decided to see the nurse.

"You're in a family way," she said, using the Victorian euphemism for pregnancy. The nurse gave Mary a few words of advice and a small pamphlet, none of which registered past the shock barrier.

The next day, Mary went to work at her usual station, testing and analyzing nitric acid shipments, her face blanche and hollow. Her manager asked if she felt okay, and she managed a "yes." At lunchtime she tried to call Patrick, but his landlady said he was out. The next weekend she bought a bus ticket to Toledo and waited for Patrick to get out of class. They took one of their hand-holding walking dates, and she told him of her condition. At hearing the news he let go of her hand, said a few things she could not fully grasp, and walked away.

That was the last time she ever saw him.

Despite growing up in a large family, Mary had always felt alone in the world—a testament to the fact that loneliness is not cured by crowds. Now, deceived by insincerity, duped by counterfeit promises, and impregnated by the sperm cells of deception, Mary was more alone than ever. She tried to assuage her anger, her sorrow, her guilt, by redoubling her dedication to work. She volunteered for extra hours, but the nausea got worse, and after two weeks she had to reduce her schedule to less than full-time.

"Will I be sent away?"

Several months had passed since Mary began wearing loose-fitting dresses. But things were steadily progressing, and fat-girl dresses could only take you so far.

"Will I be sent away?" she asked again.

The nurse seemed puzzled by the question. "Sent away? Sent away where?"

"You know. Because of my baby."

The nurse smiled. "How old are you?"

"Twenty-two." At that moment, for some odd reason, Mary wondered where she would be when she turned twenty-eight.

"Twenty-two. So you're an emancipated adult. Who would or could send you away?"

"You don't understand," said Mary. "Where I come from you're not an adult until you get married and start having children."

"Well," said the nurse. "You're halfway there."

At DeSales College Mary had attained something she had longed for from her earliest memory: independence. But that independence had come with a price: the acceptance of a $500 scholarship from the Parent Teachers Association back in Ray, plus an incentive scholarship from the Sisters of Notre Dame. Dropping out of school to assist with the war effort was one thing, but if the PTA and the nuns heard this latest news, they might want their money back. If Aunt Ida kicked her out of her house over it, Mary might be forced by financial necessity to return to the family farm—a new form of hell.

The nurse set down her stethoscope.

"When girls your age get sent away, sometimes it's by their parents. Sometimes it's by their pastor. But do you know who frequently does the sending?"

Mary shook her head.

"The girls themselves. Many of the young women your age who get sent away to wait out their pregnancies send themselves away. They don't want to face the judgment of their friends, their families, their coworkers. If you get sent away at your age, it will be because you did the sending."

Mary asked, "How often do girls do that?"

The nurse thought that over, then said, "A lot. Especially with the Jews and Catholics. Are you Catholic?"

Mary nodded, and the nurse made a few notes on her clipboard.

"It's interesting you would bring this up. I had a patient in here last week with the same problem as you. She was preparing to take a bus to an aunt's house in Texas to have her baby. She said something kind of funny—called herself a member of the "Sent-Away Club.""

That's when Mary realized she hadn't seen Kathleen McNulty since Friday.

The nurse handed her a prescription. "Come back next month and we'll see how you're doing."

—\/\/\/—

A month before, the protective jumpsuit required for her job had gone from too large to too small. It had become too tight and Mary was forced to requisition a larger size. Already several of the other girls had begun making remarks, pointed and anointed with innuendo. Mary had been working at Plum Brook Ordnance for about a year. She had already discussed with her supervisor that she might need to take a leave of absence soon. He had tried to dissuade her, of course, since Mary Sherman had quickly become one of his best chemical analysts. She would be hard to replace, even on a temporary basis. When Mary refused to give him a reason, the reason became apparent, and so he agreed.

When her due date was less than a month away, Mr. Hollingsworth insisted the time had come for her to take a leave of absence to have her baby.

"I have to know something," he said. "I have to know whether you will be back."

"Of course I'll be back," she said. "Why wouldn't I?"

"How are you, as a single mother, going to raise a child and still put in ten to twelve hours a day here at the plant?"

For family Mary had only Aunt Ida, and they did not get along well. Mary realized that if she were to raise a child, she would have to give up her job. If she gave up her job, she would have no means of support. Without any means of support, she would have to return home—to the Sherman farm.

Never. I will never live there again.

And so Mary began the process of giving her child up for adoption. She sought help from a local Catholic charity, who referred her to a church-run hospital in Philadelphia. Mary arranged with Plum Brook for a two-month leave of absence, then spent a little money and purchased her first decent suitcase. As she prepared for the move to Philadelphia, she could not get her mind off of how dreadfully she had strayed from the course she had set for herself back in Ray. This pregnancy was not supposed to happen—this child was not her destiny.

On the morning she was to leave, Mary penned a letter of gratitude to her aunt, making it clear she would return in two months. At the end of the letter she wrote: *Please don't tell my parents.*

This time when she boarded the bus she sat in the back seat. As the bus started forward, Mary turned around and waved a tearful good-bye to an empty parking lot, watching with wet eyes as Toledo, Ohio, vanished into the horizon.

—◦–ᴠᴠᴠ–◦—

Why? Why would my mother travel 550 miles from Toledo, Ohio, to St. Vincent's Hospital in Philadelphia to have her baby? She would have passed dozens of perfectly well-suited hospitals along the way.

I continue my online searching and finally uncover two solid explanations.

It turns out that St. Vincent's Hospital was not just any hospital; it was a hospital owned and operated by Catholic Social Services, and it specialized in providing medical and adoption services to unwed mothers. My online search stumbles across dozens of blogs and posts filled with the longings of children grown to adulthood who were born at, and adopted from, St. Vincent's. In these blogs many of them plead for help in locating their birth parents. In other blogs, birth parents plead for help locating the children they gave up.

One such post comes from a woman named Joan who is seeking her birth mother. The woman not only shares what little information she has of her mother but also speaks directly to her. As she addresses her biological mother, she explains that she loves her because, although "Mom" didn't raise her, she gave her life, a gift for which she will always be grateful.[2]

The more I read, the more I realize St. Vincent's was some sort of birth-to-adoption factory. A post dated June 5, 2009, fills in some details about the hospital, where young women from across the nation would come to birth their babies. The post provides the location as being in southwest Philadelphia, specifically at Sixty-Ninth and Woodland Avenue. The blogger explains that the orphanage was at the same location, and that eventually the hospital was closed before a new wing, called St. Vincent's Home for Unwed Mothers, was constructed.[3]

So, based on the information I garnered from various adoption websites, it would not be unusual in 1945 for a young, near-term Catholic girl to travel 550 miles from Toledo to Philadelphia to give birth. St. Vincent's was a Catholic hospital where unwed mothers-to-be gravitated from across the country.

⸺◦〰◦⸺

For no apparent reason, Mary awoke. Her eyes opened and she was surprised at how fully alert she was. Her bed was next to the window, and she pulled back the curtain to look outside. It was dark, with a gentle moon. She reached over to the end table and grabbed her watch. It was 2:00 a.m.

Mary turned her body, sweeping her legs out and down. Pushing with both hands, she managed to sit up. She took a few deep breaths, placing both hands on her parturient belly. It would not be long now; her due date was only three days away. A wall clock ticked softly, and she checked it to make sure her watch was accurate. It was. She felt an urge to urinate and carefully forced herself to stand up.

That's when the first contraction started.

⸺◦〰◦⸺

My mother, as discussed elsewhere in this book, was an expert at covering her tracks and blotting out the history of her life prior to getting married in 1951. She probably had no idea that a handwritten letter she wrote to the adoptive mother, Mary Hibbard, would be carefully preserved all these decades. Since the letter is the only record of my mother giving birth back in 1944, outside of the birth certificate, I have decided to publish the letter here.[4]

Dear Mary:

Am writing this letter—one for the public, and this one, which is restricted.

Instead of coming right home from the hospital, I stayed there for three weeks after I got up, working for my keep, and the baby's care. It saved me quite a bit and it was better than being alone with nothing to do, as I worked with a bunch of girls, all there for the same reason.[5] A lot of them are really O.K.

The baby is now at Mrs. Coyle's, and I am to go back to work Monday. I would have stayed at the hospital longer but I thought Vernon was coming so I came home. He isn't here yet, darn him.[6]

I got my insurance without any difficulty $206.50 altogether, which ain't hay. I would have been dead broke by the time I got back to work if I didn't get it.

I went to see Sister Assumpta as soon as I got your letter. She said they couldn't handle the case, but the thing to do is to take the baby over to you, then you get an attorney to arrange things there. I can bring the baby over there any weekend, but would like to know about a week in advance when you want it, so I can make plans accordingly. Perhaps you'd better consult an attorney before I come.

Mrs. Rau gave me some clothes for the baby (second hand) but they're mostly for an older baby. Some of them have the label of a shop in Glenview, Illinois, on them, which is the right direction, and far enough from Philly.

I went back for my final checkup Monday and the doctor says everything's all right. I feel O.K. too. They made me stay in bed fifteen days and take about a handful of sulfa pills every few hours because I had a temperature. And I weigh exactly 111 pounds again and I'm afraid I always will.

At first I thought the baby was chubby but after I saw the rest of her, I saw she's quite thin. However, she gained 1 pound, 4 oz. the first six weeks, she takes all her formula, and keeps it down, and she doesn't have any bone trouble. Mrs. Coyle says she's awfully strong. She put her on [her] stomach in the crib when she was seven weeks old and somehow or other she got way up into the corner. By the way, today's her birthday—she's eight weeks old.

I was just interrupted by a call from Al for a date tonight.

Be seeing ya',
Love,
Mary

10.
HIDDEN FORTRESS

"Secrets are made to be found out with time."
—WIDELY ATTRIBUTED TO CHARLES SANFORD

T he convoy of American soldiers stopped their Jeeps and other vehicles alongside the slimy mud track that, according to the map, was a road. Most of the men were on foot, having marched 175 miles in nine days. They were members of the 104th Timberwolf Army Infantry Division of the 145th Regiment stationed in central Germany in an area known as Lippstadt. The men desperately needed rest, but since resting was not permitted, their sergeant ordered a "maintenance stop" for the Jeeps—which their orders did allow.[1] The Jeep drivers opened the hoods of their cars, the mechanics made a show of working on them, and the rest of the soldiers did their best to find a soft rock or dry space on which to sit and take a breather.

Medics attended to the wounded, of which there were many.

One of those wounded was Private John M. Galione, a lanky Italian soldier who had grown up on a farm on Long Island. Pulling his pant leg up, he examined the gash in his leg where a piece of shrapnel from an exploding shell had ripped into his shin bone.[2] The medics had patched him up, but the wound kept opening and bleeding—the result of having to march more than 15 miles every day. The boot extended just above the wound, and the leather rubbed on it incessantly.

It was spring in the alpine forests of Germany's Harz mountain range. Much of the ground was covered in two to three feet of snow, several inches of which had fallen the previous evening. Deep snow cups, together with an abundance of wet mud, gave evidence that the spring thaw was well under way. The date was April 4, 1945.

Like everyone else in his unit, Galione kept a firm grip on his rifle. Although German soldiers were surrendering in droves, most of the country was still an active war zone. Adolf Hitler's suicide, and Germany's capitulation, were less than a month away, but that was a future still just a little too far away to see. Pockets of Third Reich loyalists were everywhere, steadfastly holding out for a last-minute miracle.

Time passed, and then the sergeant ordered the convoy to move on. Private Galione stood up and joined two of his buddies as they resumed their march northward. Where were they going? No one seemed to know for sure. Just keep walking, and engage the enemy when you find them.

Eventually the road petered out and the men found themselves moving cross-country with map and compass. The enormous size of the Harz mountain range, combined with the density of the thick conifer forest, triggered a thinning of the ranks. The Jeeps began having trouble finding room to maneuver and, one by one, the men and equipment of Galione's unit became spread out and separated. Their sergeant's voice could occasionally be heard resonating through the forest, ordering them to stay together, but there were too many trees, too many rolling snowfields, too many burbling creeks. Another hour of hiking through boot-sucking mud passed, then Private Galione noticed a snow berm that did not look natural, and he stepped over to take a look. Climbing the berm, he found himself standing about ten feet above a set of railroad tracks. The previous night's snow had been cleared, making the steel rails plainly visible. From this Galione suspected he had stumbled upon an active route. He was about to slide down the far side of the berm and inspect the tracks when he heard the rumble of an approaching engine. Standing downwind, he could smell the train's cargo before it pulled into view. The odor was all too familiar to Galione and soldiers like him. It was the same stinging redolence that had permeated thousands of battlefields for thousands of years. A single whiff and the genetic predisposition programmed in by a million years of evolution would make one turn their head and back away in instinctive disgust.

The pungent stench of death.

Private Galione hid behind a tree, and moments later a locomotive, trailing several boxcars, appeared. The young soldier knew from prior briefings exactly what he was seeing.

"A prisoner-transport train," he whispered to himself. He could not call out to his fellow soldiers, lest the train's personnel be made aware of their presence.

As the engine and boxcars—their Nazi swastikas boldly portrayed—passed his position, Private John Galione watched, and realized his error. At one time it had without a doubt been a prisoner-transport train, but as the tsunami of stench washed over the forest in the train's wake, John realized the boxcars were probably carrying nothing more than corpses.

When the last car disappeared around a bend, Galione slid down the berm and stood on the tracks.

"Sarge—I found something!" he called out. "Hey, Sarge!"

Galione stood quiet for a few moments, but heard no reply. He opened his field map and pinpointed his position. Oddly, there was no railroad track marked anywhere in the area. Yet there it was, directly beneath his leather boots. According to the map, nothing significant should exist there or near: no buildings, no industry, no military installations, and certainly no labor or concentration camps. And yet there was no mistaking the smell of the train's cargo. The discovery of this active, yet unknown, rail line fascinated the young soldier, and he contemplated his next move. The tracks were on a slight incline. Downhill—the direction the train had gone—was to his left. Numerous US Army units were encamped that direction, and not far away; the train and its crew would soon be discovered and captured. To his right, the tracks tantalizingly rose into the dark, snowy forest. Since the tracks were not supposed to be here, he could not help but wonder what exactly they led to. John contemplated following the tracks, but the sun was going down—it would soon be too dark for such a dangerous expedition.

"What are you doin'?"

John looked up to see one of his friends from the unit standing atop the snow berm.

"Found some tracks that aren't on the map."

"Sergeant says to make camp."

That evening John sat with the small group of soldiers he had grown close to. These were the men who had watched each other's backs in battle, who had saved each other's lives more than once. They would be forever bonded in a friendship more spiritual than casual. They sat in a circle, as had become their custom, so every direction had at least one pair of eyes on watch. As the men quietly enjoyed their meager meal, John's thoughts kept returning to the railroad track. He could not get the thought out of his mind; he kept envisioning himself alone, following the tracks to wherever they led. A trainload of the dead could mean several things, but his gut told him there must be a concentration camp in the area. There are people out there somewhere who need liberating, he told himself. People who are starving, dying, and most certainly being tortured. They needed his help. John Galione had not come to Germany to kill people, though many a German soldier had been unfortunate enough to find himself in John's gunsight. No, he had come here to save people—to save lives. He looked from man to man in their close-knit circle, each face bringing forth a memory of some blood-spattered battle, some tearful comrade slowly fading from life. John could not count how many times he had looked into a soldier's fading stare and heard him say, "Tell my mother I love her." He trusted these men beyond measure, and normally he would share his thoughts with them. But not tonight—something was different this night.

John used a cigarette lighter to check his watch: 9:00—the men would be turning in soon.

After washing and stowing his mess kit, John carefully checked his gear. Before dinner he had asked the sergeant for permission to break off from the unit and spend a few hours following the tracks. His argument was that the stench from the boxcars made it pretty clear that a concentration camp might be in the area. The officer immediately turned him down, telling John that it was just too dangerous for one soldier to be alone in an active war zone. John suggested he could get a few volunteers to go with him, but again the sergeant turned him down.

"Our unit stays together," the sergeant had insisted. "That's an order."

A light snow was falling at 9:30 as the soldiers began crawling into their sleeping bags. When he was sure most of them were asleep, John gathered his gear and shouldered his pack. To be certain that no one would think he had gone AWOL, he left word about his plan with one of his trusted comrades who was still awake. Then he marched off toward the tracks.

John had no choice but to find his way in the dark. His flashlight was useless—all the unit's batteries had expired over a week ago, and they had yet to be resupplied. Fortunately, traveling in the dark was one of the skills the 104th Timberwolf had been trained for.

It took almost half an hour of blind bushwhacking, but he finally managed to find the snow berm and the railroad track. One of his sergeant's concerns was that John would get lost in unfamiliar terrain all by himself, and then the sergeant would have to invest time and men in searching for him. But Private John Galione wasn't concerned about getting lost or being separated from his unit; he had complete faith in his army training, his compass, his map, his instinct. Besides, how could anyone really get lost when they had a railroad track beneath their feet?

The snow flurries of the past few hours remained untouched on the steel rails, evidence that no more trains had passed through. John's thoughts were interrupted by a faraway sound, and he turned to listen. Somewhere a woodpecker was doing its tree-punching duty, creating a *rat-a-tat-tat* sound that many soldiers found unnerving. Sound played tricks in deep forest cover, and sometimes it was difficult to tell the difference between a woodpecker making his home, or some not too distant German soldier mowing down soldiers with his Mauser-Werke MG-42 machine gun.[3] *I hate that sound.*

Standing in the middle of the tracks, John turned to face the uphill direction. As he took several deep breaths, the ice-cold air sent a chill from his lungs to his skin. He wanted a cigarette, but it was out of the question; he was in the open and the red embers would easily betray his position on such a dark night. Enemy soldiers were everywhere—a day never went by that his unit didn't encounter some.

John considered for a few moments his moral dilemma; he was about to disobey orders to, ostensibly, save lives. His commanding officer had made it very clear that he was not to go off on this mission, let alone all by himself. Disobeying direct orders was not something John Galione was known for. Since the first day of boot camp he had been the epitome of a good, obedient soldier. Yet here he was, in violation. He could not tell the others the reason for his disobedience. He could not tell them exactly what he was feeling. He could not tell them that from the moment he had smelled death that afternoon, a strange, inexplicable force had been driving him. Something he had never before felt. Urgency, like an invisible energy, seemed to be pushing on his body. Was it God, or just a gut feeling? Regardless, if he was not in camp come sunup there would be consequences.

I'll follow the tracks for a couple of hours, then come back. Sarge won't even know I was gone.

John checked his watch: 10:08 p.m. The moon would be up soon; he needed to take advantage of the dark. His plan was to walk as far as he could by midnight. If he found nothing, he would turn around and return to camp long before anyone had woken up. Mentally, he would chalk the whole thing up to "reconnaissance"—a common field duty for army grunts like him.

Private John Galione took another deep breath, put one boot in front of the other, and started walking.

—⋏⋏⋎⋏—

"If officers from the Gestapo, or Himmler's SS, show up at your door and tell you they need to take you somewhere for your protection, that's when you know you're about to be assassinated."

Those were the words of one of von Braun's close associates, spoken more than a year prior. Now those were the words that went through

Wernher's mind as he stood in his doorway, facing two officers representing SS Brigadier General Hans Kammler.[4] They were insisting that Wernher accompany them to an undisclosed location "for his personal safety and security." The American and Russian armies were closing in, they said, and the general was concerned that von Braun might be captured or killed.

In the past few weeks, owing to Russian army advancements toward Peenemünde, Wernher and his men had been working day and night to move their V-2 components and equipment to a more secure underground location. Deep inside a mountain range, far from popular Allied targets, General Kammler had used forty thousand slave laborers from the nearby Dora concentration camp to dig a series of tunnels and caves out of solid rock. It was there that Wernher von Braun and his rocket engineers, along with another forty thousand slave laborers, had been relocated in order to resume V-2 production. Despite frequent Allied-aircraft bombing raids, the characteristically efficient German Engineering Machine, coupled with an almost-unlimited supply of concentration-camp workers, had succeeded in reassembling the V-2 factory in record time.

No sooner had the underground factory resumed V-2 production, however, than Kammler's SS guards had appeared at von Braun's door. Wernher, always the good talker, tried to stall.

"Isn't General Kammler interested in securing the safety of my colleagues as well? We have hundreds of scientists, engineers, and technicians. Allow me some time to . . ."

"You are to come alone."

"Alone? But what about the others? Surely the general cares about their safety as well."

"Our instructions are to bring only you."

"Fine. At least allow me some time to pack a couple of suitcases."

"No need for that," they insisted.

As Wernher put on his coat, he wondered what he might have done to anger General Kammler.

Probably nothing. One did not have to do anything special to earn a bullet to the head from the general. The last time he and Wernher had been together was a memory Wernher preferred to forget. Only three weeks had gone by, but the image would be tattooed onto his memory forever. Wernher had been seated beside General Kammler in a Daimler-Benz G4 during the evacuation. As he turned around for one last view of his beloved Peenemünde rocket base, Wernher could see many of the buildings were going up in flames. Once all the equipment and V-2 parts that were worth keeping had been removed, the German military wasted no time in burning and destroying everything that was left. Hitler was proud of the advancements his country had made in rocketry and aerospace, and he had no desire to let anything fall into the hands of his enemies. Better to destroy it than to let them have any part. Days later, Wernher began hearing rumors that all "unnecessary personnel" at Peenemünde had been shot by the SS to prevent their knowledge from falling into the hands of the advancing Russian army. Now two SS officers stood before him, demanding to escort him to "safety." Stepping outside, he noticed the car they had arrived in was a Daimler-Benz G4. Wernher checked his watch: it was 2:00 a.m.

—WW—

Private John Galione had promised himself he would only walk until midnight, then turn around. But when midnight arrived, and the tracks had not led him to anything significant, he decided to keep going.

Just a little farther.

Whenever there was a curve in the tracks, John would convince himself that whatever he was looking for—whatever the tracks were leading him to—was "right around that bend." But then he would groan or exhale with frustration as each turn produced the same result: more track, another bend, more disappointment.

Just a little farther. Just a little farther.

After a while, John stopped to catch his breath. Though the incline was gentle, the track was relentlessly uphill. He checked his watch. It was 2:00 a.m.

For the past hour, the moon had been playing peek-a-boo behind puffy, swift-moving clouds. When the moon was out, he could see several hundred feet ahead. When it was obscured, the night would become as dark and lonely as a tomb. Sometimes he could not even see his feet. During those moments of intense pitch, John would guide himself by sliding his rifle barrel along one of the steel rails.

For now, the moon was out and his visibility was good. John looked down the track, examining the way he had come. Then he looked up the track, hoping to spot some clue that his journey was at an end. Downhill was food, water, safety, a warm fire, and the companionship of his fellow soldiers. Uphill was danger, the enemy, the unknown, and death. Moral dilemmas were a daily occurrence during wartime, and Private John Galione had certainly experienced his share since arriving in Europe. Nothing quite like this, though. On the one hand, he had been well trained to be a disciplined solder, to follow orders. Besides, everyone knew the war was almost over—why take any unnecessary chances when the troops would soon be going home? On the other hand, there could be hundreds of people at some concentration camp up ahead who needed help. They could be starving, dying. If he were one of them, wouldn't he want to be rescued? John was sure of only one thing: if his insubordinate sojourn was to go undiscovered, he would have to turn around now.

Just a little farther.

No, I should turn back.

A small cloud moved in front of the moon, and all went dark.

John faced the uphill direction, trying to get his boots to move. He kept weighing the pros and cons, his mind doing battle with his conscience. A few quiet moments passed, then suddenly someone pushed him hard from behind, almost knocking him over. As he was straightening himself up, he was shoved forward again, even harder.

Instinctively, John turned around, raised his rifle, and chambered a bullet. He held the butt to his shoulder and pointed at the total blackness behind him.

"Who's there! Answer or I'll fire!"

At that moment, the cloud moved away from the moon. Dim lunar light poured over the railroad tracks and the surrounding alpine forest. There was no one.

—⟋⟍⟋⟍—

During the late-night drive, von Braun kept wondering what a bullet to the brain felt like, or if one would even feel such a bullet at all. Attempting to relieve his anxiety, he had chatted up the two SS guards seated on either side of him, but they had said little. If he was being escorted to his death, they gave no indication.

Hours later, the car arrived at the office of General Kammler. The two SS guards escorted von Braun, then stood on either side of him as they waited. A few anxious minutes passed, then the general entered the room and sat behind his large desk. He got right down to business.

"Dr. von Braun. We both know the war is lost—it's only a matter of time. What we have to do now is prepare for what will happen next. Germany is about to be overrun by its enemies. Each of us has to ensure we aren't hanged as a result."

Von Braun nodded quietly, still uncertain where the general was going with the conversation.

"Here is what you will do: you will make a list of your five hundred very best men. We will put them on a train to a secure location under the protection of the SS. No families will be permitted. There you and your men may continue your rocketry work. The place is very isolated, so it is doubtful our enemies will find it for some time."[5]

When von Braun discovered that General Kammler had no intention of shooting him and had indeed arranged for his safety and security, he was surprised. He returned home and began preparing Kammler's requested list, wondering what fate awaited those who would be left off it. Von Braun also wondered if his own name would be removed from the list if Kammler got wind of what he had just asked his friend Dieter to do. Most likely, von Braun would be summarily executed.

—✧✧✧—

In a few years, Dieter Huzel would live a comfortable sun-drenched upper-middle-class life in a suburb of Los Angeles called Woodland Hills.[6] He would work for North American Aviation in the same department as one of the company's hot new propellant analysts, Mary Sherman. But that was still several years into a murky future, a future where the prospects for survival vacillated with every passing moment.

Dieter downshifted the truck's transmission into second gear. The gradient of the mountain road had steepened, and the engine was slowing toward a stall. He checked his rearview mirror to make sure the second truck, driven by fellow engineer Bernhard Tessman, was keeping up. On March 12, General Kammler had ordered the burning of all charts, books, and records pertaining to rocketry work in Germany.[7] Von Braun, with the help of his longtime supporter Walter Dornberger, recently promoted to general, decided to disobey the SS officer's order and secretly arranged for a convoy of trucks to move the materials to an abandoned mine in the Harz Mountains. It was that convoy that Huzel now led, a convoy that made its way past numerous military check-points using counterfeit orders and paperwork. Huzel, Tessman, and a handful of other engineers and hired laborers were essentially risking their lives to save technology.[8] For von Braun and Huzel it was personal; they had spent a large part of their lives helping to invent this technology, and they could not bring themselves to either surrender it or destroy it.

And so, when von Braun asked if Huzel could undertake an effort to secret away all their blueprints and records in an old mine, Huzel readily agreed.

Now all they had to do was make sure the Gestapo and the SS never found out.

—✧✧✧—

Mary was sound asleep, having had a very tiring day at work. There was a time not long ago when she would awake full of stamina and energy, but that was before. Sleep came easy to her these days, and not just because her body demanded it. She could sleep because she believed in what she was doing—making the world safe for democracy, one gunpowder magazine at a time. She and her fellow TNT-making sisters were so committed to the war effort that many of them spent as much as 10 percent of their meager earnings to buy war bonds.[9] But that commitment displayed itself even more in the physical sacrifices they endured: harsh conditions, demanding workloads, and laboring long hours in unheated, uninsulated buildings. The structures were so poorly put together that sometimes snow would fall *inside*.[10] One day one of Mary's coworkers, June Franklin, received a letter informing her that her husband had been wounded in battle in North Africa. Putting down the letter, June immediately marched into the payroll office to buy another war bond. Her job was to nail the bottoms onto the wooden crates that the TNT was packed and shipped in. Promising her fellow coworkers that she would never again miss a minute of work, she resumed hammering, proudly signing her name to the bottom of each box—a personal gift to the American soldiers.[11]

There seemed no end to the insatiable thirst of the army and navy for gunpowder and other explosives, and Plum Brook Ordnance was their chief supplier. For three and a half years, Plum Book operated twenty-four hours a day, seven days a week.[12] It was a factory whose products were designed to separate enemy hands from arms, legs from torsos, heads from bodies. They made nothing less than the highest quality wartime explosives ever developed, and the thousands of women who worked there were proud of it. It was they who had helped make Plum Brook the number-one supplier of gunpowder, and one of the top three companies producing TNT, in the United States.

Plum Brook did not achieve this milestone without help. They enjoyed the kind of partner every entrepreneur would love to have in their hip pocket: the United States government. Using federal assistance, the founders of Plum Brook forcibly took the land of 150 family

farmers, crushing their dreams under the tank treads of wartime necessity.[13] The farmers were paid for their land, of course, but at amounts
far less than market value. Rather than argue, all of the farmers
moved on to farms and employment elsewhere, with one notable exception. Fred C. Baum, a third-generation farmer and lifelong Sandusky
resident, chose to fight. The government had offered him $18,375 for
his property. He felt it was worth much more, and he sued. A judge
agreed with Baum, awarding him $31,700.[14] No one will ever know
what would have happened had the other 149 farmers chosen litigation
over capitulation.

The construction of Plum Brook had been achieved with such haste
that the builders estimated the provisional structures had a life expec-
tancy of only five years.[15] Once the plant was up and running, Plum
Brook's output soon averaged 400,000 pounds of explosives per day. Its
construction proved serendipitous. Ordnance production began on the
morning of November 15, 1941, just 22 days before the Japanese would
launch their surprise attack on Pearl Harbor.

By 1945, Plum Brook had produced almost one billion pounds of
explosives.[16] The enemies of democracy were taking major hits and were
in full retreat in some areas. But they were hardly going quietly into
the night. Fighting in both Asia and Europe continued to be furious.

From the 500-pound bombs dropped on Peenemünde, to the spit
of gunpowder in every one of Private Galione's bullets, the munitions
made at Mary's factory were crucial in fueling any hope of an Allied
victory in Germany.

—◦〜〜◦—

John Galione was being fired upon.

Utterly alone, and out of contact with any Allied combat units, he
had been walking almost nonstop for five days and nights. As the sun
rose on the sixth day, Galione found himself standing fifty yards from
the entrance to a large man-made tunnel. Directly in front of it, parked
on a rail siding, were several boxcars, some empty, some filled with

corpses. As John began examining the bodies in one of the cars to see if any of them were wearing the uniform of American or Allied soldiers, he accidently dropped his rifle magazine, and the sound, as slight as it was, drew the attention of a lone German guard.

John took shelter behind a large pile of boulders as the guard ran toward him, firing one bullet after another. John fired back. Despite the fact that neither man had good shelter, none of the bullets hit their targets.[17] Eventually the lone guard simply ran away. Of that moment, John would later write, "I don't think the Germans wanted to die this late in the war either."[18]

When he decided it was safe, John began to patrol the area, especially the tunnel entrance. Without knowing how many Germans, if any, were inside, he chose not to explore the tunnel deeply. But inside he could smell the stinging odor of some kind of munitions. Because of the possibility that the tunnel might be booby-trapped, he went back outside and surveyed the area in front. He saw a road and decided to walk down it a short ways. Rounding a curve, he came upon a high barbed-wire fence with a heavy, padlocked gate. Not far away, inside the gated area, he saw several men he took to be prisoners. The men stared back at him with hollow eyes and pale expressions.

Both the tunnel and the locked gate would turn out to be major discoveries.

This lonely journey by a single, determined army private striking out on his own, changed the course of history. After five days and nights alone in enemy territory in the German alpine wilderness, Private John Galione had not only stumbled upon the underground hiding place of the V-2 rockets and components, but the infamous Dora concentration camp as well (which some historians would later consider more notorious than Auschwitz).[19]

Galione immediately began retracing his steps to rejoin his unit. In an amazing stroke of luck, he soon came across two American soldiers who were making minor repairs to their Jeep. They gave him a lift, and within hours, Galione was notifying his superiors of what he had found. Colonel Holger N. Toftoy wasted no time in sending a fleet of trucks to Mittelwerk. Once inside the tunnel, army technicians managed to collect enough components to build 100 V-2 rockets, many of which were eventually flown at the White Sands Proving Grounds.[20]

—•/\/\/\•—

Within a few days of von Braun's meeting with General Kammler, SS officers and guards began rounding up all five hundred of the men on von Braun's list. At each home they visited, one gut-wrenching experience after another played out like some theatrical drama as the engineers and technicians were literally pulled from the grasps of their wives and children. No one knew for sure their destination or destiny. All they knew, from long experience and a thousand stories, was that if the SS showed up at your door to forcibly remove someone, the results were often deadly. Though the scientists had no way of knowing it at the time, most of them would survive, thanks largely to a clever subterfuge that would later be cooked up by none other than their esteemed blond-haired, blue-eyed leader.

Von Braun was sitting up front and was the first to see the tall metal gate. Owing to recent surgeries to repair a broken arm and shoulder suffered in an automobile accident, von Braun was one of the few allowed to arrive by car. Most of the other five hundred men would be arriving on the "Vengeance Express," a train that had been dedicated a year before for the sole purpose and use of transporting the rocket engineers.[21] A sentry, seeing the official SS Mercedes approach, did not even ask for identification and opened the gate wide. The vehicle entered the compound without even slowing down.

Over the next few days, von Braun and his men took up residence at what officially was referred to as a Wehrmacht camp. Located near Oberammergau, the mountain hideout had been used as a retreat for high-ranking German officers.[22] Because of this, the accommodations were better than what many of von Braun's men had enjoyed in their own homes. In the words of von Braun, "The scenery was magnificent. The quarters were plush. There was only one hitch—our camp was surrounded by barbed wire."[23] But the facility also had a hidden purpose. Those in the know called it the "Alpine Redoubt," and it had been built as part of a contingency plan in the event Germany was overrun by its enemies. Here in the remote forested foothills of the Alps, Hitler would come to command his troops in one final, magnificent battle that would vanquish their enemies and preserve the Third Reich.[24] At least that was the grand theory.

It did not take long for the German scientists to figure out General Kammler's real plan. They had been told they would be able to continue their research, yet none of the tools they needed to perform that research had been provided. General Kammler was under the impression those tools—the charts, records, and books—had all been destroyed. Only von Braun and Dornberger knew of Huzel's clandestine mission to preserve their work and legacy, and they certainly weren't telling the SS general. The relocated scientists were highly intelligent individuals—some of the best minds on the planet. They were smart enough to realize something other than research was on the mind of their keeper—that what they were really being held for was to ensure not their own safety and security, but the safety and security of General Kammler. The general had gathered Wernher von Braun and his engineers together like a stack of poker chips—something to bargain with in the event the Allied armies captured the general and threatened to hang him.[25] The men also knew that if Kammler's gambit failed, they would all be shot in order to prevent their knowledge from falling into the hands of Germany's enemies and, of course, eliminate witnesses to Nazi war crimes.

But General Kammler was not the only one who knew how to play a poker hand. He would soon discover the hard way that any individual who attempts to outthink five hundred rocket scientists is likely to be the player who loses his chips. Von Braun and his men managed to escape the luxurious prison General Kammler had put them in using a ruse so simple it is hard to believe it worked. One day, a number of Allied bombers dropped their explosive cargo on a target nearby. Von Braun told the SS officer in charge, known as the Sturmbannfuher, that if the Alpine Redoubt were bombed Germany would lose all of its great rocket technology in minutes. He convinced the officer that the best solution was to disperse his men to private homes in towns around the area. That way, the logic went, some of them might die, but many would survive. Unable to communicate with Kammler, the Sturmbannfuher agreed, and just like that, von Braun managed to talk himself and five hundred men out of prison.[26]

—◦◦◦◦◦—

On the morning of May 8, 1945, Mary awoke at her aunt's apartment to the sound of a hundred car horns. She looked at her bedside clock: 7:10 a.m. There was a rumble of a dozen fast-moving feet on the hall stairway, and she heard someone yelling. Mary stepped to the window and pulled back the curtains. Two stories below, the streets were filling with people, many of them still in their nightclothes. They jammed the sidewalks and mobbed the streets, preventing any semblance of traffic flow. Music was playing from somewhere, crowds were dancing, everyone was cheering, strangers were kissing.

She opened the window and the first words she heard were, "Germany has surrendered!"

Mary leaned out the window and shouted, "What about Japan?" She repeated her question two more times, but the clamor of the growing crowd was too great; no one could hear her. She turned on her radio and tuned the receiver for news. It did not take long—it seemed every station had information. It was true: Germany had officially surrendered. But the fighting in the Pacific was raging as hot as ever.

Hundreds of thousands of young American women were waiting for their boyfriends and husbands to come home. Those who had loved ones in Europe would no doubt be crying and screaming in unrestrained happiness as soon as they heard. Those with loved ones in the Pacific would congratulate the first group, then pray that the new peace would spread. Mary, who was waiting for no one, would simply collect her last paycheck. Mary knew a two-tier social system was about to be created—those who knew their men were coming home, and those who had to wait. It would dramatically change the environment at work. Even so, it was obvious the war could not last much longer. Two of the three major Axis powers had fallen—it would be impossible for an isolated island nation like Japan to hold up against all the combined Allied powers that were now free to focus on a single theater of battle. The end seemed near.

Mary's first thought was the future of her friends and coworkers at Plum Brook Ordnance. As she watched the feral revelry fill the neighborhood, Mary wondered how many of the Plum Brook women understood that they were about to receive pink slips. Most of her coworkers would be so euphoric over the anticipation of seeing their loved ones again, they would be blind to the approach of their new peacetime reality: joblessness. No war meant no business. No business meant no employees. Mary knew that once a peace treaty was signed by Japan, employment at munitions factories like Plum Brook Ordnance would vanish into the ether. The city of Sandusky, which depended so much on munitions contracts, might vanish as well. Just so much chaff on a windy Ohio day.

Mary had anticipated this moment. Only fools thought war-related employment lasted forever. Three months before, when news of troop successes in Europe was being reported daily, she had bought an old typewriter at a pawn shop. Now, pulling it out from beneath her bed and grabbing several sheets of paper, Mary Genevieve Sherman began to craft her post-war résumé.

—◦—◦—◦—

A week prior to Mary crafting that résumé, von Braun, Dornberger, and several of their associates had hidden themselves in a hotel high up a mountain near the German–Austria border. On May 1, the day after Hitler had committed suicide, von Braun was in an upbeat mood.

"Hitler is dead, and the hotel service is excellent!"[27]

He and his engineers needed an emissary—someone who would be willing to leave the relative safety of their numerous hiding places and go out into the world, looking for American soldiers. They had voted and, almost to a man, decided that surrender to the Americans was highly preferable to falling into the hands of Russia's brutal and unforgiving Red Army. Owing to his passable command of the English language, Wernher's younger brother Magnus was chosen.[28] On the morning of May 2, Magnus hopped on a bicycle and steered himself down the steep road on the Austrian side of the border.[29] After several hours, Magnus returned, saying that he had met some American officers from the CIC (Counter Intelligence Corps), but they did not believe his story that he knew where the inventors of the V-2 were staying.

"They want some of you to come back with me so they can question you."[30]

Wernher, Magnus, and Dornberger then took four others, piled into three staff cars, and drove down the mountain to meet the Americans. During the drive, all seven experienced various levels of trepidation. What would happen to them? Would they be interrogated? Beaten? Tortured? Sent to prison?

According to von Braun, "When we reached the CIC I wasn't kicked in the teeth or anything. [Instead] they immediately fried us some eggs."[31]

On September 2, 1945, the Empire of Japan capitulated and sent representatives to a treaty signing aboard the battleship USS *Missouri*. Around the world, millions of people celebrated what they assumed would be a new world full of peace and harmony among nations.

They were wrong.

11.

A NEW KIND OF WAR

"The average defender operates in a fog of uncertainty."
—WIDELY ATTRIBUTED TO HUGH KELSEY, BRIDGE CHAMPION

I look in the mirror to make sure I'm ready. Hair combed, glasses cleaned, gray slacks, white shirt, dark blue blazer. It's the uniform of Catholic school, St. Catherine of Sienna, Reseda, California. The year is 1964, and I'm in fifth grade. There is no middle or junior high in their system; we attend first through eighth grades at a single school. I gather my notebook, well-organized with ruler, paper, number-2 pencils, and a fountain pen. Ballpoint pens are popular now, but they are absolutely not allowed at St. Catherine's. No one, of course explains why they are not allowed. Rules are rules, and that's just all there is to it.

I step out of the bathroom and walk to the dining room. My mother is there, playing four-handed bridge by herself. She has the morning paper, a mug of coffee, and she's smoking a cigarette.

"Bye, mom," I say. "I'm going to school now."

I wait a few seconds for a reply, but there is nothing. She says nothing, does nothing, acknowledges nothing. I'd give anything for a little bit of something, but everything is nothing.

The only time my mother grants me any attention is when I do something wrong. If I drop a plate and break it, she will shower me with attention, but it won't involve hugging and kissing.

"Bye," I say again, then turn and walk out the front door.

Through the open front window I can hear her gather the cards, tap them into a neat pile, and begin shuffling.

For years this has been my routine: get up by myself, make breakfast by myself, shower myself, dress myself, go to school by myself. I live in a very myself world. Of course, I'm certainly old enough to do these things on my own, but I've been doing them much longer than most kids. I'm like one of those latchkey children, except my mother is always home.

I leave the house at 7752 Lindley Avenue and begin my lonely journey to school. St. Catherine's is only four blocks away, but it's a very dangerous four blocks. In 1964, the cliché of milk-money-stealing bullies is not yet a cliché; it's just my life. And those who walk alone are the easiest targets. I've complained to my mother about it many times. She responds by reshuffling the deck.

There are certain defensive tactics I've developed over the years. Pretending I'm sick and staying home from school is my favorite. When that doesn't work, I leave fifteen minutes early and get to school before the milk-money goons station themselves at their usual street corners or hiding places. Bullies, I've learned, don't like to get up early. Like lions on the Serengeti, they prefer the antelopes that straggle. So I learn early on not to be a straggler.

When you walk alone day after day you have a lot of time to think. Think, and fantasize. I do a lot of fantasizing, imagining my mother as a loving person who kisses and hugs her son before he goes off to school. I fantasize she is someone who shows affection and isn't afraid to say "I love you" once in a while. Sometimes I really stretch things and imagine she makes my breakfast or (*gasp*) drives me to school so I can spend time with her (and not be victimized by the Guidos).

Like all good fantasies, they never happen.

Something is wrong with my mother, but I don't know what it is. She's unhappy, but whatever it is that's bothering her she refuses to talk about. Sometimes she gets angry for no reason and lashes out. I rarely see her cry, but it happens. The terrible thing is I just assume all mothers act this way—I have little to compare her to. That assumption, however, is called into question each morning when I arrive at the front gate of St. Catherine's and see how many mothers are hugging their children good-bye. They even kiss them. They talk to them. Everyone smiles and looks happy.

Why is my mother so different? Decades later I would ask my father that question. His response made perfect sense in light of what I had experienced: "Your mother loves babies," he said. "She just doesn't like children."[1] That alone, of course, did not explain many of her behaviors.

I've taken a circuitous route to school this morning, eluding the Guidos by zigzagging through a retail section of town. As I enter the front gate of St. Catherine's, I try not to watch the huggy-kissy mothers seeing their kids off. I don't want to look, it's just too painful. I hang around one of the tetherball poles until the bell rings. Two hundred blue blazers and two hundred blue plaid dresses line up by grade level for the morning bugle. I can still hear the tune in my head, but I don't know the name. I think it something Sousa wrote for the Marines.

When the prerecorded bugle music is over, we watch the raising of the stars and stripes, then we're dismissed to classes.

A couple of hours later, I look up from my times-tables exercise to check the clock. It's Friday, 9:59 a.m. The clocks in St. Catherine's are always well calibrated, and if they aren't, you can always readjust them every Friday morning. In one minute, a siren somewhere in the neighborhood will begin its low wail, slowly building in both tone and intensity. It will reach a shrill peak, then slowly fade. This buildup/fade-out wail will continue up and down for about a minute, during which all of us will be on our knees and under our desks, our hands clasped firmly over our heads. This is the drill—the "duck and cover" drill that has become ubiquitous throughout the American school system, both public and private. Every Friday morning at precisely 10:00 a.m. we, the children of the United States, practice a safety procedure that, we are told, will keep us safe from a nuclear holocaust. No one ever articulates any specific details about what a "nuclear holocaust" is or what effect it would have on us, but we suspect it must be something really bad. All we've been told is that if we just get under our desks, all will be fine. We don't mind, of course, as it's a fun diversion from the routine of our otherwise-boring school day. So every Friday, we do the drill.

But not the teachers. All the classes are taught by habit-wearing nuns, and each one of them, we've decided, must have some kind of death wish; they never participate in the drills. As the students duck and cover, the nuns casually sit at their desks, grading papers or reading a book. My teacher prefers crochet. The religious order these nuns belong to requires them to take on a masculine name. My teacher's name is Sister Robert Francis.

I raise my hand and she calls on me.

"You know what I would do if I were the Russians?" I ask.

"No, George," she says. "What would you do?"

"I would time my attack so it occurred at ten on a Friday morning so everyone would think it was just a drill."

I mean it as humor, but no one is laughing. Instead my comment is met with silence—silence that is interrupted by the low moan of the city's civil-defense siren whirring up. As we all duck under our desks, I notice Sister Robert Francis, for the first time, is participating. Her ankle-length black habit quickly disappears under her desk.

After a minute, the siren fades out, and we climb back into our desks to return to the scholastic grind. Nobody has the courage to tell our teacher she is now covered in dust.

Years later I would discover that the danger we practiced duck-and-covering drills for was not just an atom bomb, but an atom bomb carried aboard an ICBM—Intercontinental Ballistic Missile. I would also discover that my mother had been instrumental in creating the very thing that made us cower under our desks.

The hours pass like a dull headache. Finally 3:00 arrives and we're dismissed. As large crowds of my fellow students greet their mothers at the gate, I head home the same direction I arrived: alone.

—⋁⋁⋁—

Irving Kanarek was an American citizen, born and raised in New York. But his parents were immigrants, having been born in the Soviet Union. In 1947, Irving decided to apply for a chemical-engineering position with North American Aviation, then one of the country's leaders in the fledgling business of liquid-propellant rocket design. NAA had been advertising for scientists and engineers relentlessly in engineering trade journals—the boom times had arrived and they needed bodies with brains.

Irving showed up at the company's main office, where a dozen other men, all in dark suits, white shirts, and crew cuts were sitting at tables, filling out applications. A young man missing his left arm handed him a sheaf of papers, then directed him to a table. Irving could smell the atmosphere of wartime veterans—the room was filled with it. Most of these men, he was certain, had been in Europe or in the Pacific two years earlier, fighting the Axis powers. Now the war was over, and a million men needed real employment.

As Irving filled out the application, he arrived at a small box on page 2 that one of his friends had warned him about. The question beside the box read, "In what country were your parents born?" In the post-war atmosphere of anticommunist hysteria that had gripped the United States like a vise, many applicants would have lied; they would have written anything but the truth. To admit that one's roots could be traced back a single generation to America's new Cold War enemy would be considered employment suicide. Irving's friend had strongly encouraged him to write "Poland" instead—a country whose forced takeover by the Nazis, followed only a few years later by a Soviet Red Army invasion, was a country whose citizens enjoyed a great deal of sympathy in America. When fanning the flames of anticommunism, American politicians would use Poland as an example of why all good Americans should fear and despise the Soviet Union. But Irving chose to be honest. His parents, he argued, had not been born in one of the Soviet satellite countries, but in the Soviet Union itself. He was an engineer, and engineers cared about facts. Fudging facts for expediency was what politicians and criminals did, not practitioners of science. And so, on his employment application, in the little box requesting the birth country of his parents, Irving Kanarek put pen to paper and clearly wrote the well-known abbreviation for the Union of Soviet Socialist Republic: USSR.[2]

Irving completed the remainder of the application, affixed his signature, and dropped it into a large box overflowing with the efforts of a thousand other applicants. Then he left.

The soldiers and officers, veterans and vagabonds lining up at the North American Aviation main office weren't the only ones looking for post-war work. They had returned from the theater of battle with promises of full employment—promises that frequently went unfulfilled. The year was 1947, and the country was weighed down by a glut of available workers. For this reason, almost all of the women who had been employed in wartime work—Rosie-the-Riveters et al.—had been forced out of their posts in order to make room for the massive influx of jobless male veterans.

Some, though, did not go so quietly into the night. One such rebel was the little unwashed urchin from North Dakota.

The Slauson bus pulled to a stop near the intersection of a busy street with no street sign. Stepping down to the sidewalk, Mary Sherman adjusted her cat's-eye glasses and squinted at the bright Southern California sun. Everyone who had regaled her with stories of how much better than Toledo the Los Angeles weather was had been truthful. This was a city with weather so perpetually perfect its citizens joked about LA having only two seasons: spring and two weeks of winter.

The intersection was only three blocks from Los Angeles Airport,[3] and Mary briefly watched as a small twin-engine plane left the runway and ascended over the Pacific Ocean. Six feet from the bus stop bench was a high security fence topped with a menacing helping of barbed wire. From the corner she could follow the gate in two directions—one traveled east, the other south. The chain link seemed to go on forever in either direction, and the "front gate" she had been directed to enter was nowhere to be seen. She flipped a mental coin and started walking south. After two minutes, and still no gate, she decided to ask for directions. An elderly man walking a dog was approaching.

"Excuse me, sir. I'm looking for the front gate to North American Aviation. Do you know where it is?"

He pointed and gesticulated wildly at the fence, then walked on.

Mary shook her head and kept walking. A minute later she came to the compound's southwest corner and another intersection. She turned eastward, continuing to follow the fence. Halfway down she came to a locked gate marked "DELIVERIES." She waved at a nearby guard, who stepped over.

"I'm looking for the front gate."

"Are you a visitor?" he asked.

"I'm here to apply for a job."

"Go to the next corner and turn left. It's down about a hundred yards."

"Thank you."

As she continued east, the hum of heavy machinery whirring like a generator could be heard from inside the closest building. It was the sound of people making things, she thought. Finally, after four years of designing explosives and bombs she would be able to make things that were useful.

If she got the job.

Turning right at the northeast corner, she could see a line of vehicles turning into a large entrance. Mary approached the gate and turned in, a lone pedestrian amongst a fleet of Detroit-born sedans. A guard waved her over, and she explained the purpose of her visit. The guard pointed her in the direction of a large glass door on the far side of the parking lot.

●──◇◇◇──●

Irving pushed on the metal handle of the door, but it did not budge. He pushed again, but nothing happened. Then he noticed the "PULL" sign above the handle. Obeying the written directive, he pulled, and the door opened without effort. As Irving walked across the hot asphalt pavement toward his '44 Ford, he held a hand over his eyes to blunt the bright morning sun. Someone was coming toward him.

A young woman.

Irving guessed she was about twenty-five years of age. Her dress was conservative yet attractive, but there was something off about her clothing—something not exactly normal. Her dark hair ended just above her shoulders, and she wore cat's-eye glasses. As they neared each other, he noticed she had a small mole near the left of her mouth—what women had started calling a "beauty mark." They passed each other without comment, and Irving removed his car keys from his pocket.

It was not until fifteen minutes later, as he was driving down Imperial Boulevard, that Irving realized what was "off" about the woman's dress: the creases in her collar and the folds in the sleeves did not have the slick look of machine-tooled precision one saw in department-store clothing. Irving realized the woman's dress had been hand sewn. Either by choice or by necessity, the woman, whoever she was, preferred to make her own clothes.

The next day, an NAA Human Resources secretary assigned to make cursory inspections of all new applicant paperwork pulled Irving Kanarek's application from the pile on her desk. As the initial reader of all prospective employee applications, one of her responsibilities was to use a bold black pen to make spelling corrections. At the top of the form she noticed this applicant had misspelled "engineer," and she corrected it. The rest of the page seemed fine, but on page 2 she came to the parental-birth-country box and saw the letters USSR. She had never seen that abbreviation on any of the hundreds of applications that had come across her desk—it was completely alien to her. Occasionally she would see *France*, *Mexico* or even *Germany*, but for the most part foreign countries had become rare. More than 95 percent of the time, the box she was now staring at contained the same predictable entry: USA. Perhaps the young woman never read the newspapers. Perhaps she did not own a television. But for whatever reason, the secretary was unfamiliar with the abbreviation USSR, and assumed it was a mistake. Using her bold black pen, she "corrected" the entry, changing it to read USA.[4]

The rest of the application appeared to conform to company standards, and she tossed it into the "OUT" bin, where it would be closely examined by various department managers trolling for talent.

•—\/\/\/—•

The instructions above the door handle read "PUSH." Mary pushed the door open and stood before a single desk.

"I'm here to apply for a job."

The secretary pointed her toward a door, and Mary walked through it.

The day Mary Sherman walked into the Employment Services office of North American Aviation, she was struck by how things had changed in just two years. During the war, almost all of her coworkers, and many of her supervisors, had been women. As she took her place in line to obtain an employment application, there were eight men ahead of her, and several more lining up behind. At the tables were at least twenty prospective employees filling out job applications—all men. Looking around the room, she felt for the first time like a fish out of water. This was the world of crew cuts, black slacks, white shirts, cheap ties, and roving eyes. Things had changed a great deal since her days at Plum Brook Ordnance. Mary had not been worried about obtaining employment—until this moment. She carried with her numerous letters of recommendation from Plum Brook, each more glowing in their praise than the next. Still, this was an engineering job she was applying for, and the want ad had specified "college degree preferred." She could hear the men talking in low tones, sharing stories of their war wounds, their exploits, their medals, their college degrees obtained under the GI Bill. She had no such benefit—few women did. In the post-war world, women were expected to vacate the jobs they had occupied while American men were overseas so employment space could be freed up for the over one million returning American GI's. Now here she was, a poor farm girl from North Dakota with no college degree, competing against a system designed to promote the careers of men and return women to the kitchen.

But Mary Sherman would have none of it—she would not be blocked, she would not be stifled. Mary stiffened her resolve, threw back her shoulders, stood as tall as she could in her five-foot-five, 111-pound frame, and turned toward the man sitting at the table.

"I'd like an application for one of the chemical-engineering jobs you advertised."

The man was unsure what to do. Of the thousands of engineering applicants who had stood before him, none of them had been women. Finally he asked a question.

"Do you have a college degree?"

"No, but I have work experience, and many references."

"References from whom?"

"Plum Brook Ordnance."

The man's expression was blank.

"You worked at Plum Brook."

"Yes, sir."

"As what, exactly."

"I was a chemist. I worked on distilling various solutions of weapons-grade dinitrotolulene. We were the main supplier of . . ."

The man held up his hand.

"I was a corporal—served in the 6th Army under General Walter Kreuger. The munitions we received from Plum Brook were always first-rate. Your company saved a lot of American lives, and helped win the war."

He pulled out an application and handed it to her.

"Pens are on the tables. Fill it out, sign it, drop it off over there. Good luck. Next."

As Mary walked to an empty table, she could feel everyone in the room watching her.

━━∿∿∿━━

Tap, tap, tap . . .

Tom Meyers nervously tapped the eraser end of his pencil on the glass cover of his desk. One of several supervisors in the Engineering Department of the Office of Research and Development, Tom had been given the responsibility of hiring 130 more engineers and analysts. His sphere of responsibility focused on the engineering of new liquid-fuel rocket engines and their propellants. His assignment was to "push the envelope"—to create ever more powerful rocket engines that used ever more powerful propellants so the government could achieve ever more powerful performance. He had a great many positions that needed filling, and filling them was finally starting to get easier. The quality of the prospective employee pool had improved greatly over the last six months. The first crop of post-war college graduates was out there, looking for jobs and searching for the American Dream. That large group of applicants comingled with an even larger number of 4-F's who had sat out the war years getting their own degrees. Even so, there were several key positions that remained vacant, and none of the applications coming across his desk stood out.

Especially problematic were the three open positions for theoretical performance specialist. The job of a TPS was to mathematically calculate the expected performance of a particular rocket-propellant combination when used with a particular engine. In this way his department could make a great many engineering decisions without having to go through the time and expense of actual testing. Field testing of rocket engines and their propellants was expensive and time consuming; better to math it out whenever possible.

That morning, Tom had filled two of the TPS positions—good men with excellent education and experience. But the candidate pool with the right qualifications was shrinking. Other supervisors in other departments had already gone through the same pile of applications, each one taking the best and leaving the rest. And the more supervisors who slogged through the application pool, the more it degraded into a slush pile. Even so, he had managed to hire a number of very good people. And he did it by looking for small details that other supervisors tended to overlook. A paper résumé, he knew, was a highly imperfect way to judge a flesh-and-blood human being.

Ideally the job of a theoretical performance specialist required someone with superior skills in both math and chemistry. Hands-on experience with rocket propellants or exotic chemicals was a plus. But every application he looked at had the same problem: If someone had the right experience, they lacked the math skills; if they had the math skills, they lacked the chemistry; if they had the chemistry, they lacked the real-world experience.

Then he came to an application from a young woman, Mary Sherman. Superb math skills. Excellent knowledge of chemistry. A great deal of real-world experience. Still, there were two major problems with the applicant: she did not have a college degree, and she was a she.

Tap, tap, tap . . .

An 8 × 10 family portrait adorned one wall of Tom's office. Ten years ago his wife had given up a promising career in order to become a full-time wife and mother. She tried to keep her feelings hidden, but Tom knew she secretly harbored regrets about giving up what could have been a lucrative and promising path. She was happy, yes, but even happy people have regrets. Theirs was a good marriage, and their children were a delight. They got along well, and there were few complaints. In fact, the primary complaint his family had was Tom's propensity to nervously *tap, tap, tap* his pencil on things when he was thinking. Which was often.

Tap, tap, tap . . .

Tom set Mary Sherman's application aside for a moment and looked out onto the engineering floor. Below him, hundreds of steel and wooden desks were stacked in ranks and files, with a starched-white-shirted man at each one. Tom smiled at the mental image: nine hundred white shirts and ties surrounding a single dress. The idea was delicious, and for a brief moment he thought about hiring the young woman just for the novelty. There were other women on the floor, of course: file clerks and typists. Still, for the engineering department it would be novelty.

Tom considered for a few moments the new state of affairs that had begun to dominate people, politics, and policy. It was a war without bombs or battles—a war fought with philosophy and threats rather than guns and tanks. That's why they called it the Cold War. And the enemy was the Soviet Union. As the holder of a top secret government security clearance, Tom was privy to certain information most Americans had no access to. One such item was that the Soviet Union was well on its way to building its first atomic bomb. World War II may have ended, but the world was, in many respects, an even more dangerous place than ever. Both countries were designing large rocket systems. It was only a matter of time before someone figured out that placing an atomic warhead on an unstoppable missile was preferable to dropping it from a vulnerable bomber. This was a world so new and precarious that every morning brought another fear-engendering revelation and headline. But until interplanetary travel became a reality, everyone had to live in that world. And so this was no occasion for frivolous novelty. If he were to hire this woman, she would have to own the chops for the job.

Tom could tell by the smudges, fingerprints, and dog-eared paper that several other managers had already looked at this application, and passed on it. One of them had even written a comment at the bottom: "No way!" Still, he decided to give it a closer look. Having to sift through hundreds of job apps and résumés was a mind-numbing experience, and important details were sometimes missed. Today was no exception, for though he had checked her prior work experience, somehow he had skipped over her prior *employer*. According to the application, Mary Sherman had spent the previous four years doing chemistry work at a place he was very familiar with: Plum Brook Ordnance.

Tom Meyers stopped tapping his pencil.

12.

WHITEWASHED IN WHITE SANDS

**I saw a new life. There was nothing left for us in
Germany.**
—WALTER WEISMAN, GERMAN ROCKET SCIENTIST[1]

For almost two hundred years the United States had enjoyed its
advantageous geographical position, set apart and isolated from
humankind's traditional warring states of Europe and Asia. For all the
detrimental arguments that could be made against the philosophical
megalomania of Manifest Destiny, the push of the country westward
had created a formidable country—one that was protected on both its
east and west flanks by vast oceans. This isolation and separation was
one of the key factors that made the American Revolution possible, as
the British were forced to transport men and materials back and forth
across thousands of miles of sea from England.

This oceanic buffer zone has always been an enviable advantage for
the United States. During World War II geographically contiguous countries such as France and Poland were easily picked off by Germany due
to their close proximity and ease of access over established roadways.
The geographical advantage the Americans enjoyed was a sore spot
for Adolf Hitler, and his fantasies of world domination required him to
spend a great deal of time considering ways to overcome it. It was for this
reason that he began to envision a rocket that could fly a warhead across
the Atlantic Ocean and hit America.[2] He asked Wernher von Braun if
it might be possible to build a rocket with such a range. Von Braun,
who had spent his life dreaming of building rockets that could fly to the
moon or Mars, was quick to say yes. No one who wanted to spend money

building larger rockets ever got a "no" from Wernher von Braun. What military leaders saw in purely weaponry terms, however, he saw in terms of opportunity—an opportunity to use their money and manipulate their zeal in order to advance space-travel technology.[3] And so, long before it had an official name, Adolf Hitler envisioned the invention of yet another new weapon—the intercontinental ballistic missile, or ICBM. With such a weapon he could make moot America's geographical advantage. World domination would be within his grasp.

Germany would never build such a weapon, since the Führer and his government died before any serious work could be done. But what Adolf Hitler could only imagine, others would soon create.

From his window in the confines of the former enlisted men's barracks at Fort Bliss, Texas, Wernher von Braun watched a small dust devil swirl and dance. It traveled a short ways until reaching a cement walkway where it dissolved, its sandy particles made powerless by the lack of new fuel. This was his new world, the world of endless sand. Sand and rocks and gravel and more sand as far as one could see. The sands of southwestern Texas were so vast they made the beaches of Peenemünde look like a child's playground. And it got into everything: eyes, nostrils, mouth, numerous other bodily cavities, carburetors, air-conditioning, rugs, drinking water, food supplies, and, absolutely worst of all, the propellant tanks. The sun never set on a day that Wernher did not crunch kernels of grit between his teeth, wash granules from his eyes with great handfuls of water, or order his men to clean the propellant storage and feed systems one more time.

Then there were his colleagues. The ever-declining morale amongst the German émigrés was causing dissension. Having left their homes, families, and the unspoiled German alpine forests for the sand-blasted military barracks of Fort Bliss, 104 immigrated rocket scientists were quietly questioning their newfound loyalty to America and its people. For these men, the Texas desert was not some isolated whistle-stop on a long rail line to somewhere better—according to the US Army it was the end of their line and they would do well to get used to it.

And to make matters almost inhumane, the beer tasted like mouthwash.

But the worst sand problem was yet to come. Too far ahead to see, the day was coming when a sibling rivalry–like competition between three branches of the US military would intentionally prevent von Braun from launching the world's first man-made satellite years ahead of the Soviet Union. The US government would accomplish this by preventing von Braun from working on orbit-capable rockets, assigning him instead to ballistic-only designs. The absurdity of this waste of manpower would reach its farcical zenith with the United States eventually ordering von Braun to load the Redstone with two thousand pounds of sand as ballast to intentionally lower his rocket's performance. The reason? To make certain von Braun's Redstone did not "accidentally" get into orbit during test flights.

But that was still almost a decade away.

Having been coddled with gourmet food, luxurious conditions, and generous military funding at Peenemünde, many of the German scientists soon began pining for home—a home that no longer existed. Yet despite the wartime destruction of their country, a small number of the scientists turned their back on the United States and returned to Germany.

For those who stayed, their Fort Bliss lifestyle consisted of tasteless US Army food and endless boredom. An exciting evening was defined as scorpion hunting in the bed sheets before turning in at night. If the hostile weather and inhospitable living conditions of Fort Bliss weren't bad enough, several times a week the scientists had to endure a lengthy bus ride over rutted, jaw-jarring earthen roads, crossing state borders to reach their new work environment at New Mexico's White Sands Proving Grounds—where more sand awaited. The German scientists began referring to themselves as "prisoners of peace."[4]

For all this, the Germans had the US government's Project Paperclip to thank.[5] A technology transfer program designed to help America catch up to the Germans (and stay ahead of the Russians), Project Paperclip had quietly imported a select group of Germany's most elite rocket engineers to America. The operation was managed by the US Army and kept top secret for a very good reason—no one in the US government or the military wanted to suffer public censure for harboring or collaborating with Nazis. Of course, one could try explaining that the men were *former* Nazis, but the US Army was taking no chances—the last thing it wanted was a public-relations nightmare on its hands. In this regard, Project Paperclip had a second goal—to settle those scientists in a location so remote and inaccessible that over time their Nazi pasts could be effectively whitewashed.

Thanks to von Braun's superior public-relations talent, that goal was not only achieved, but achieved beyond anyone's expectations.

At first there was not much to do at White Sands. Traveling from their homeland, the Germans had been given high transportation priority, securing rides to the United States on returning Army Air Corps planes. This meant they preceded by weeks, even months, the delivery of the V-2 missile parts being shipped by boat, then rail, from Mittelwerk to White Sands. When the parts did finally begin trickling in—loaded into hundreds of boxcars strung out across the country— von Braun soon discovered the trans-Atlantic crossing had wreaked havoc with his precious rocket. During transit, much of the V-2 systems and components had become rusted and corroded in the salty ocean air. Many of the railroad boxcars, despite their top secret contents, became waylaid, sidelined for months on spur tracks—lost and forgotten in the mega-chaos of their transport. But the US government was salivating over the German technology and had little patience for nagging little details like lost shipments or the natural oxidation of metals. The army ordered von Braun and his men to begin assembling, fueling, and firing V-2's without delay—no excuses. When von Braun explained that many of the shipped components were no longer usable, the army accused him of stalling. In the end, it took von Braun and his engineers eight months to cobble together enough parts to begin flying V-2's on a semiregular basis.

That eight-month lag time turned out to be more crucial than anyone suspected. For though the day was not far off when the sand-blasted rocket tests and studies performed at White Sands would assist in propelling America into an undeclared space race, few realized that the race had already begun. While the von Braun group was cooling their heels in Texas and New Mexico, the Soviets were busy. The Americans had picked much of the fruit from the German rocket-scientist tree, but not all of it. There were many German engineers and technicians who had been captured by the Soviets and taken to Moscow. Immediately upon arrival, Joseph Stalin put them to work. For the next five years, those scientists would be fiercely set to their labor by the Stalin Machine. They would design and build the largest and most powerful rockets ever conceived.

Like Project Paperclip, the Russians kept their endeavors secluded and highly secret. That secrecy, however, would not last long. On an isolated patch of ground known as the 5th Tyuratam Range in a place called Kazakh, the Soviet Union was preparing a mammoth project that would spread fear to every man, woman, and child on planet Earth.

13.

ALIAS CHIEF DESIGNER

"There is no such thing as an unsolvable problem."
—WIDELY ATTRIBUTED TO SERGEI KOROLEV

Mary's attention went straight to the return address on the envelope: North American Aviation in Downey. She quickly ripped open the envelope and unfolded the letter.

> *We are pleased to inform you that you have been accepted for the position of analyst at North American Aviation. This position is located at our company facility in Inglewood. Please call our office at your earliest convenience and let us know if you intend to accept this offer of employment.*

She chose to accept the offer in person, taking the bus the next morning to the Inglewood building. It was there she met Tom Meyers for the first time.

"What is an analyst?" she asked after pleasantries had been exchanged.

"They analyze."

"Analyze what?"

"Data."

"Ah."

Only much later did she discover that the work of an analyst was not that much different from an engineer. Six months later, one of her coworkers let her in on the secret.

"An analyst is an engineer without a college degree. They call them 'analysts' so they can put them in a lower pay scale."

After arranging to renew her top secret security clearance, Mary Sherman reported for work at NAA on July 15, 1947. She was assigned to Department 95, the Aerophysics Lab. A young female clerk escorted her through the massive maze of 400-pound, gray steel desks, all arranged in orderly ranks and files, in the largest room Mary had ever seen. Everywhere she looked, there were young white-shirted men with ties and crew cuts.

Throughout the building, cigarette and pipe smoke had accumulated into a large cloud, hanging almost motionless in the air like an LA summer inversion layer. An avid smoker of Winstons and Kents, Mary had no objection.

After a minute of walking, the clerk stopped and pointed to one of the steel desks.

"This will be yours. That's your file cabinet. The employee lounge and bathrooms are over there. Slide rules, reference books, everything you need are in the drawers. Any questions?"

Mary nodded. "I do have one question. What do I do next?"

"Next, you sit down and wait for Mr. Meyers."

"Mr. Meyers. The man who hired me?"

"One and the same. He's your boss." The girl was about to leave, then turned back.

"We wanted to congratulate you," she said.

"For what?"

"You're the first woman to be hired on the engineering floor. It's a big day around here for us girls. We have a special table in the lounge we have lunch at. I know you'll usually eat with the engineers, but I hope you can join us once in a while."

Then she turned and walked away.

Of course I'll join you—why wouldn't I?

Mary executed a playful 360-degree swivel in the chair, making a mental note at how little office furniture had changed in the past ten years. The heavy-duty metal desk, the gray-leather swivel chair with four wheels, the drab gray file cabinet with the chrome handles—all of it identical to what she had worked with at Plum Brook. Due to wartime metal rationing, all new desks at Plum Brook had been made of wood, but there were still hundreds of the pre-war steel ones scattered throughout the complex. Now that two years had passed since the armistice, everyone was moving back to the 400-pound, gray steel behemoths. Somewhere in the world there was someone in a major decision-making position whose concept of a proper office desk was too much steel and plenty of gray paint.

"You must be Mary."

She swiveled around to face a man leaning against her file cabinet. He took a sip from a bone-white coffee mug, holding out his free hand to shake hers.

"I'm Carl Amenhoff. That's my desk right over there."

He pointed to a desk across the aisle. Mary stood up and shook his hand.

"Nice to meet you, Mr. Amenhoff."

"No, no—we're all on a first name basis here. Just call me Carl."

"What do you do here?"

"I'm a TPS."

"TPS? I'm sorry . . ."

"Theoretical performance specialist—just like you."

Her eyes took a split second to dart back and forth from his left hand. He was married.

"I'm a TPS?" Mary couldn't help but smile. She had never had a job that could be melted down to an abbreviation.

"That's what you were hired for."

"I see. And what exactly is a theoretical performance specialist?"

"Don't ask him—he doesn't know."

Carl and Mary turned to see the arrival of a man dressed in a dark suit and starched, white, long-sleeve shirt. He seemed to be dressed far too warmly for Southern California.

He's married, too.

"Don't listen to him," said Carl. "This guy doesn't know squat."

"I know how to sign your paycheck."

"Okay fine—he knows one thing."

Both men laughed, and the anxiety that Mary had been feeling began to ebb. There was a folksy manner to these rocket scientists that she suspected would make her working environment pleasant enough.

"You'll be performing complex studies to determine specific impulse values of new propellants." He held out his hand. "I'm sure you remember me—Tom Meyers, head of the department."

They shook hands.

"So you needed someone with experience in thermochemistry, heat transfer, fluid flow dynamics, mixture ratios—that sort of thing."

"Yes," said Tom. "That's exactly what we needed. Do we have that with you?"

Mary looked at Carl. "I don't know about this guy, but you do have that with me."

Both men laughed again.

"Carl will fill you in on your assignments. Good luck." Then Tom moved on to some other 400-pound, gray-steel desk destination.

"So I'm ready to work. What do we do first?"

"Soon as Dick Mascolo and Ed Royce get back from break we'll have a sit-down and go over who does what."

"What kind of projects will we be working on?"

"Our first priority is the NALAR."

"Nalar?"

"North American Liquid Air Rocket. It's a new air-to-air attack missile we're working on. The Army Air Corps doesn't want to use solid propellants until they're more reliable. So we're working on a liquid-fuel version."

Mary glanced over at Carl's desk and noticed a copy of one of Alfred Sheinwold's bridge books.

"You play bridge?" she asked.

"A lot of the engineers play bridge. Every day, at lunch time."

"I'd like to get in a game. I'll need a partner."

"Sweetheart, you're the only girl in the engineering department. Trust me, you're not going to have any trouble finding a partner."

—/\/\/\•

The faucet was dripping.

Every eleven seconds, a turbid drop of selenium-laced water would force its way up the steel spout to the edge of its lip. It would hang there a moment, wiggle a bit as surface tension fought with gravity, then release itself at one g of acceleration, until it finally plopped into the cast-iron cooking pot thirty centimeters below.

The relentless regularity of that rhythmic plopping was distracting the Russian scientist from his work. His name was Sergei Korolev, and he had a bedeviling heat-transfer problem with his rocket's regeneratively cooled engine. The latest static test of the R-1 engine had been a spectacular failure, having blown a hole clean through the combustion chamber, eviscerating the engine, and seriously damaging the test stand. The R-1 was rated for one-hundred-eighty seconds of nonstop operation, yet could never manage more than fifty.

Plop.

Korolev pulled his fur-lined coat tight around him and buttoned it. He exhaled, and a cloud of warm vapors steamed from his mouth. He looked down at the multitude of equations, notes, and chicken scratches that filled hundreds of papers, strewn haphazardly over every square centimeter of the termite-riddled table. Every few seconds, a cold Siberian wind would push against the shack, rattling the boards and penetrating their cracks. The bitter breeze would flutter and scatter the papers, incrementing the clutter and aggravating his temper. Sometimes Korolev would allow one of those papers to float aimlessly to the dirt floor, other times he would reach out and snatch it.

Most of the engineers and technicians in Sergei's employ were less than enthusiastic about living in an environment as inhospitable as Tyuratam. But after spending six tortuous years in Kolyma, Kazan, and numerous other gulags and prison camps, Korolev had become inured to any climate that resulted in bottomed-out thermometers. In fact, he might have still been in those gulags were it not for the intercession of a number of his fellow engineers and colleagues who wrote letters and testified on his behalf. For this reason, Sergei Korolev was thankful—thankful to be free to perform the task he loved more than anything in the world: building big rockets.[1]

The scientist took a momentary break to sip from his cup of tea. He used that moment of respite to scan his surroundings. His "office" at the newly constructed Tyuratam Rocket Base was less than spartan: the table, a small shelf unit, a forty-watt light bulb, a fireplace, a window, a sink. At least he had running water—a benefit owing to his station as chief designer. And thanks to a small gas generator, he also had a few amps to power his light bulb. Plumbing and electricity—they were benefits no one else at Tyuratam enjoyed.[2]

Plop.

Sergei stood up and approached the window. The temperature difference between him and the outdoors was slight, but still enough to fog up the glass. He used the soft wool of his coat's sleeve to brush away several wide streaks of moisture, allowing him a view of the flat, treeless greenery that surrounded the shack.

The vast, open savannas of the Russian steppes were not unlike the sweeping grasslands of North Dakota. Endless, undulating plains of green, peppered here and there with the occasional tree or shrub, the steppes were too dry to support a forest, too wet to evolve into desert. In the steppes of southwestern Russia, Mary Sherman would have felt right at home (at least as far as climate and topography were concerned). Unlike North Dakota, however, the Great Steppe, as it was called, had for centuries been home to boisterous hopak-dancing Cossacks, horse-riding nomads, and brutal pirates.[3] Stretching from Turkmenistan through Uzbekistan, and on into Kazakhstan, the Russian steppes had cemented their reputation into Russian folklore. And folklore was where it ended, since very few Russians had the physical, mental, and emotional stamina required to survive on the stark, featureless grasslands. It was a three-thousand-mile expanse of unremarkable nothing—a haunting place with nary a trace of civilization. Those who did attempt to settle here rarely lasted more than a few months. Water was scarce, food had to be imported, and if you wanted a house, you had better know how to construct it yourself—but good luck finding building materials. Russian wives generally refused to live on the steppes, and if a man found himself transferred there as part of his employment, he might as well find a good divorce attorney.

Korolev's own marriage could have been a case study in unions destined for self-immolation. In 1931, he was working full-time during the day as a technician maintaining hydroplanes.[4] At night he would spend another six or more hours as an unpaid intern on a rocket-powered glider program.[5] In the midst of this chaotic schedule, Korolev took one day off to marry his high-school sweetheart, a young medical student from Odessa named Ksenia Vincentini.[6] After a few vodka toasts, Korolev put Ksenia on a train to resume her medical-school education in the Ukraine. Korolev then resumed his midnight-oil lifestyle. By 1946, the marriage would be over, the result of long, extended employment-related separations, dual super-star career success, a six-year sentence in numerous gulags and prisons for Korolev (for the crime of living under Stalin's rule), and an affair with a very young girl named Nina.

Plop.

Though life on the steppes was less than attractive, to the team of scientists looking for a place to launch the up-and-coming Russian missile program, the land held great promise. Unbroken for hundreds of miles by neither mountain range nor valley, the steppes held the advantage of allowing more reliable line-of-sight radio transmission and reception—an ideal asset for those dependent on rocket telemetry systems. And the paucity of human inhabitants would make rocket launches safer. Once Sergei's crew started launching, there would no doubt be a certain percentage of failures. It would be far better to rain down fifty tons of metal and toxic propellants onto vacant land than onto some poor Ukrainian village.

The door opened and a teenage boy, Utigur, entered with an armload of firewood.

"Chief Designer, sir. Here is the firewood you requested."

Sergei nodded toward the fireplace, and the boy carefully stacked the wood nearby. Though he enjoyed his newly exalted title, Sergei was not pleased that the army had instructed everyone to call him by it, even his closest associates. The name Sergei Korolev had been officially declared a state secret by Premier Khrushchev himself, and everyone was on strict orders never to call Sergei by anything other than "Chief Designer." The Soviet leader had made it clear that anyone responsible for leaking the Great Designer's identity would be executed.

"Chief Designer, do you need anything else?"

Sergei shook his head, and the boy left.

Sergei stepped away from the table, crouched down, and threw a few small logs into the fire. He removed a cigarette from his coat pocket and used the flames to ignite it. The cigarette lit quickly, and he placed it between his chapped lips. A breath in, a breath out. He stood up and glanced at a photograph tacked above the fireplace. It was a black-and-white photo of Wernher von Braun standing at a podium, giving a pep talk to a room full of American Boy Scouts.

Korolev owed a great deal to von Braun. The Russian army had requisitioned dozens of train-car loads of rocket components from the German scientist's rocket program. The defeated Germans had wanted to obliterate everything pertaining to their vaulted high-tech program, but in the end, the American and Russian armies had advanced too far too fast for them to complete the intended destruction.

If hardware were silver, human expertise was gold.

Far better than all the boxcar loads of V-2 components were the more than fifty German rocket scientists who had been captured and put to work for the Soviet rocket program. Many against their will, of course. Unfortunately for Sergei, the greatest assets of all—Wernher von Braun and his top echelon of engineers—had managed to elude Stalin's grasp. As a group, the best and brightest of von Braun's men had surrendered to the US Army just ahead of the Red Army's arrival. Sergei wondered if the Americans appreciated the value of their booty. He would give anything to have von Braun on his team. Or a dozen of his best engineers. Oh, the rockets they would build!

The two men had much in common. Like von Braun, Korolev had

been born with stars in his eyes, dreaming since boyhood of building a spaceship that could take him to the moon and beyond. And like his German/American counterpart, Korolev had to deal with the inevitable philosophical and dogmatic conflicts brought on by national politics. In both countries, those politics were powered by the presumptive need for greater and ever more menacing military hardware. The political superiors of both men were more interested in the weapons potential of rockets, rather than as instruments of science and peace. Each man had the same problem and, independent of each other, hatched the same secret plan. Korolev and von Braun had forged agreements with their superiors to design and build massive missile systems for each country's defense. As their leaders demanded, Korolev and von Braun would make their rockets powerful enough to carry nuclear warheads thousands of miles. But tacitly, they would also design those rockets in such a way that they could perform double duty. An intercontinental ballistic missile would be constructed so that, with minor alterations, it would also have the ability to place a satellite into low Earth orbit. These were men with vision—men who could see a future that others, blinded by fear and misunderstanding, could not. They knew the race for space was inevitable and that it was just a matter of time before human-built hardware (and humans themselves) would be sent aloft to circle the Earth. And yet this knowledge did them little good in 1950. Sergei and Wernher were up against the same problem: they needed funding, and the only sources of that funding were politicians suffering the paranoia and suspicions left over from the post–World War II world.

Plop.

Sergei took a last puff of his half-finished cigarette, then carefully snubbed it out on the fireplace bricks. He set it aside for the morning; in the remote Russian steppes where supply deliveries were highly unreliable, nothing was wasted. Not even half a cigarette.

Sergei opened a military-surplus tool kit and removed a small, corroded pipe wrench.

14.

RED

"True love is like a pair of socks: you gotta have two and they've gotta match."
—WIDELY ATTRIBUTED TO ERICH FROMM

The dark green Chevy drove around the parking lot looking for a space. It finally found one at a far corner where two barbed-wire chain-link fences came together. The car maneuvered perfectly between the lines and its engine cut off. The driver's door opened, and a smartly dressed young man stepped out. He wore a conservative suit and tie, accented with a pair of flip-down sunglasses. The most striking aspect of his appearance was his hair. It was a bright red-orange color, a hirsute replication of youthful autumns spent in Vermont, and, like a beacon, could be clearly discerned from two hundred yards away. It was cut in a conservative, military crew-cut style, a World War II holdover that had become the new fashion chic for anyone working in any aspect of the aerospace business. At Caltech his fellow students, and pretty much everyone else, had called him "Red"—to such an extent that most of those students, and his professors, had long since forgotten his given name. He preferred his middle name over his first name, and always requested people use that instead: Richard. But the nickname Red just seemed to stick to him, like sidewalk gum on a hot Pasadena afternoon. Eventually Red got used to his signature sobriquet and gave up registering any objection. What he didn't suspect, however, was that his colorful appellation was about to follow him to this, his first post-university job.

It was 7:30 a.m. and warming quickly. A native Southern

Californian, Red was well accustomed to the bright morning sun that peered over the Hollywood Hills to the north. He took a glance at the buildings of North American Aviation two hundred parking slots away, then pushed the flip-up sunglasses into the down position.

It was September 1950.

Red opened the Chevy's trunk and pulled from it a medium-size cardboard box. It contained the tools of his new trade: a chemical reference book, books on metallurgy, heat transfer, and thermodynamics, two new cartons of yellow number-2 pencils, a pencil sharpener, a large eraser, a protractor and ruler, several slide rules, some graph paper, and an 8 × 10 framed photograph of the Caltech Beavers 1946 football team. He held the box with one hand and closed the trunk with the other.

Red walked with a slight limp, the result of a knee injury suffered during a game against Occidental College two years before. His position was left tackle, and in the third quarter of the last game he took his position a little too seriously, jamming his knee hard into the helmet of a falling opponent. For Red, that would be the end of Caltech football (a few years later, Caltech football itself would disappear). Even so, he had made his mark at a school that would one day be named the world's top engineering university.[1] That mark, however, would not be made in biology, chemistry, nuclear physics, or any field of academia. Rather, Red would have his name and memory forever etched into Caltech lore by breaking the law. It was during his freshman year that the Pasadena City police cited Red for carrying not one, not two, but *seven* passengers down Colorado Boulevard on his 45-cubic-inch, WWII-surplus Harley-Davidson motorcycle.

He paid an eight-dollar fine and thereafter got invited to a lot more parties.

As Red continued his walk toward the large complex of buildings, he noticed his shoes were acting tacky—as if there was something sticky on his soles. He looked down and realized the asphalt was freshly laid and had not yet lost its wet-slurry adhesive quality. He resumed his walk and soon arrived at what looked like an entrance. A guard shack and two security guards stood in his way. A dozen other employees were lined up, displaying their identity badges and awaiting permission to enter. The line moved fast, and in a minute Red found himself showing off his crisp new badge with his shiny, smiling portrait. The guards eyed the new recruit suspiciously (suspicion being the most important part of their job). The name on the badge did not read "Red," of course—government and corporate secrecy regulations actually required legal birth names.

"George Richard Morgan," said the guard, reading the name tag. The guard checked the name on a list, then waved him through.

Three years later, "Red" would become my father.

By the end of 1950, chemistry had advanced to the point that most of the viable fuel and oxidizer compounds had been theoretically calculated, and many of them had been synthesized. Each compound had its particular advantages and disadvantages—it was an axiom that there was no such thing as a "perfect" fuel or oxidizer. There were all sorts of plusses and minuses that had to be dealt with. Liquid oxygen (LOX), for example, was an excellent oxidizer, but to keep it in a liquid state it had to be kept chilled to just under –297° Fahrenheit, which presented numerous storage and handling challenges.[2] Hydrazine makes an excellent fuel, and unlike LOX it is a liquid at room temperature, but hydrazine is highly toxic and extremely unstable (i.e., explosive).[3]

Once all the best oxidizers and fuels had been theoretically invented, physically created, and materially tested, rocket-propellant experts moved on to a second phase of new propellant creation: the "cocktail." A cocktail was a mixture of two or more rocket propellants for the express purpose of obtaining better results than each compound could provide individually. Like so many other innovations, the rocket-propellant cocktail had been invented by the Germans. In World War II, Wernher von Braun had used a fuel mixture of 75 percent ethyl alcohol and 25 percent water for the V-2.[4] This decreased the rocket's performance but allowed the engine to run cooler, increasing its reliability. This "trade-off" concept, whereby you give up something you want in order to gain something you need, governed much of rocket-propellant science in the early days, and still does.

On the day Richard Morgan walked into North American Aviation, Mary Sherman was working on just such a cocktail. It was a mixture of two oxidizers: oxygen and fluorine.[5] Fluorine was nothing less than the very best oxidizer in the universe. Unfortunately, the characteristic that gave it that honor was also its drawback; it reacted with almost everything, making transport, storage, and handling extremely dangerous. Fluorine was utterly unforgiving—make a mistake, and people die. But its potential as a rocket propellant was so attractive that Mary decided to try to "tame" fluorine by mixing it with a lesser oxidizer. Since the two chemicals were miscible (they could mix in a single container without reacting with each other) a fluorine/oxygen cocktail offered some intriguing performance possibilities. She started calling it FLOX.

On this day, Mary was working on the theoretical performance of a 50/50 mixture. She had been leaning over her Friden calculator all morning, punching endless numbers into it, and her neck muscles were aching. She looked up and stretched those muscles for a moment, and noticed the red crew cut walking down an adjacent aisle. The fact that he was carrying a cardboard box into the building, instead of the other way around, meant he was a new recruit.

"Who's that guy?" she asked, turning to Irving.

Mary and Irving watched as one of the clerks intercepted the red crew cut at a T-intersection of aisles and led him to an empty wooden desk about a hundred feet away.

"No idea," said Irving. "But he looks barely out of college. They just keep getting younger and younger around here."

At lunchtime, Mary stopped at Tom Meyers's table before going through the cafeteria line.

"Tom—who's the redhead we just hired?"

"Oh yeah. Richard Morgan. He's our new hotshot heat-transfer specialist. Pretty smart guy—Caltech graduate."

Mary nodded, then turned to get her lunch.

"Oh, hey," added Tom. "You'd probably like him. I hear he's a great bridge player."

Cute, funny, handsome, intelligent, single—all great attributes to be sure. But if a guy could play bridge, then relationship possibilities had to be considered. Mary was past the age most women got married, and she was sure she was at least three or four years older than the new redheaded recruit. But bridge was a force like fluorine: not easily tamed.

"Can you set something up?" she asked.

Tom smiled and gave her a wink. "What are great managers for?"

Six months later, Mary Sherman became Mary Sherman Morgan.

Above: Mary Sherman's high-school graduation photo, May 1940. *Courtesy of G. Richard Morgan.*

Right: Richard and Mary Morgan's wedding day, July 29, 1951. *Courtesy of G. Richard Morgan.*

Mary was interested in many facets of science, including geology. Here she is holding a quartz crystal she dug up in the Mojave Desert. *Courtesy of G. Richard Morgan.*

Richard and Mary with their firstborn son and future biographer, George. *Courtesy of G. Richard Morgan.*

Burro-ing through the Yosemite wilderness. The author and his younger brother, Stephen, ride tandem. *Courtesy of G. Richard Morgan.*

Mary (*center*) holds her first-place bridge trophy. One of many. The standing onlooker is her proud husband, Richard. *Courtesy of G. Richard Morgan.*

Mary at her desk at North American Aviation. Note the mechanical calculator at the right. *Courtesy of G. Richard Morgan.*

One of the Redstone main engines powered by hydyne. *Courtesy of Rocketdyne.*

Redstone engine undergoing a horizontal hot-fire test at the Santa Susana Field Laboratory. *Courtesy of Rocketdyne.*

The Redstone/Jupiter C rocket with *Explorer 1* satellite leaves the launchpad, powered by liquid oxygen and hydyne, January 31, 1958. *Courtesy of NASA.*

Left: A local paper, the *Huntsville Times*, proudly announces the achievement of America's first satellite. *Courtesy of NASA/Marshall Space Flight Center.*

Below: The *Explorer 1* post-launch press conference included some theatrics. William Pickering *(left)*, James van Allen *(center)*, and Wernher von Braun *(right)*. *Courtesy of NASA.*

Left: The author with Irving Kanarek. *Courtesy of the author.*

Below: Bill Webber (*left*) and Bill Vietinghoff (*right*). *Courtesy of the author.*

Scene from the November 2008 Caltech production of *Rocket Girl*. *Left to right:* David Seal (Tom Meyers), Cliff Chang (Bill Webber), Hui Ying Wen (various roles), Jon Napolitano (Joe Friedman), Meg Rosenberg (various roles), Christina Kondos (Mary Sherman Morgan), Kevin Welch (Irving Kanarek), Garrett Lewis (Don Jenkins), and Todd Brun (Colonel Wilkins). *Courtesy of the author.*

Left to right: Caltech president Jean-Lou Chameau; G. Richard Morgan; the actor who played him—Jim Carnesky; Bill Webber; the actor who played him—Cliff Chang; and the director, Brian Brophy. *Courtesy of the author.*

15.

POLITICS, PHILOSOPHY, TELEVISION, AND *CUSH' SOBASH'YA*

"Rockets are large, rockets are small,
If you get a good one, give us a call."
—Daniel G. Mazur, manager,
Vanguard Operations Group[1]

I n America, Wernher von Braun had two things the Soviet Union could never offer Sergei Korolev: *Collier's* and Walt Disney.

In the early 1950s, von Braun wrote a series of articles for *Collier's Weekly* about the possibility of human space travel and what it would take to achieve it. With a circulation of four million, the *Collier's* articles made a deep impact on the consciousness of the country, galvanizing the public's imagination in the same way von Braun had been influenced as a young man by the novels of Jules Verne.[2] The attention of the American public became riveted on the very real possibility that Verne's outlandish stories might become science fact. It had already happened with the invention of Verne's submarine, why not rockets to the moon?

On the strength of the *Collier's* articles, Walt Disney asked von Braun to make some appearances on his new television show, *Man in Space*.[3] For Disney there were very real commercial reasons for this, including his desire to promote Tomorrowland in the recently opened Disneyland.[4] But Walt Disney had personal reasons as well, not the least of which were the boyhood fantasies that had imagined Disneyland in the first place. Walt Disney, like many Americans, was excited about the future of space travel. The addition of von Braun's appearances on the show describing what human space travel would be like soon had the imaginations of the American public firing on all cylinders. Thus began a public-relations campaign on the part of von Braun to get the country philosophically attuned to his frequency: that human space travel was our destiny, and since the technology already existed to achieve it, we ought to get started.

All this von Braun–centric publicity created a quiet breech in the nation's space focus—a breech so subtle almost no one was aware of it at first. On the one hand, the US government had begun work designing large rockets that might one day be capable of manned flight, while on the other hand, they had assigned von Braun to rocket projects intended as non-orbital weapons. The situation was like telling the quarterback of the championship football team he was not allowed into the stadium.

Several years later, when Vanguard rockets were failing and exploding, the philosophical breech in American rocket policy would suddenly lurch into the open. As Eisenhower and his minions scratched their heads over why the people wanted their heads on a platter, the stadium roared for its quarterback. It was one of the first illustrations of how television, though still in its infancy, would come to mold public policy—the policymakers be damned. All of that was yet in the future. For now, Disney's weekly television show helped Americans, and many people around the world, come to know Wernher von Braun as a warm, friendly neighbor. As Madison Avenue wonks might say, Wernher von Braun became a household name.[5]

And through all of this, Sergei Korolev was full of envy. The Soviet politburo had chosen to keep his identity a well-guarded state secret. No speeches, no public appearances, and certainly no television. Wernher von Braun was quickly attaining that which Korolev lacked but always wanted: respect, fame, hero worship. This imbalance in the way the world treated him and his German/American counterpart angered Korolev, but it also steeled his resolve. Where von Braun failed, he promised himself, he would succeed. Russian intelligence had ascertained that von Braun's talents were being frittered away by the

Americans. Apparently the US political machine had exiled the man to some dusty outpost known as Fort Bliss and given him almost no responsibility or authority over space policy at all.

And so Korolev began selling his ideas to every comrade and politician who would listen, including Khrushchev himself. While the Americans fiddled over the burning of Rome, Sergei Korolev would prove to the world that it was he, not von Braun, who deserved their adulation. The former leader of Germany's advanced rocket program could make cutesy television appearances, perform gooey speeches before ladies' clubs, and continue to have his ruggedly handsome mug plastered over half the magazine covers in America. It was all just a pile of *cush' sobash'ya* as far as Sergei was concerned.

Sergei Korolev resolved to earn his place in history without kowtowing to such artificial media god-making. Sergei Korolev would waste no time, squander no moment. He would show the world its mistake in idolizing the former German Nazi, and he would do it with a project so imposing everyone would have no choice but to pay attention. Sergei Korolev would build the largest, most powerful rocket anyone had ever conceived.

Taking a sip from a jet-black cup of coffee, Sergei manipulated his slide rule and performed his next calculation.

YOUR VERY BEST MAN

**"No one will ever win the battle of the sexes; there's
too much fraternizing with the enemy."**
—HENRY KISSINGER

E rnst Stuhlinger was one of von Braun's most experienced and
trusted engineers. He had been an integral member of the German
A-4 and V-2 production teams, providing numerous crucial solutions to
design problems as they inevitably arose day by day. Now, as he made
the long trek from Hangar 3 across the tarmac toward von Braun's office,
he stared at the mathematical symbols and formulas before him, and
they stared back. For Stuhlinger and the other engineers, math was
a beautifully perfect science with whom no one could argue. If a math
formula said something was true, then it was true. No matter where you
were in the universe, two plus two would always equal four, seven would
always be a prime number, and force would always equal mass times
acceleration. Mathematics offered the infinite surety and permanence so
lacking in all other aspects of human endeavor. Math. Lovely, sexy math.

But perfection has a downside; it oftentimes carries bad news.
For though it was an axiom that "if a math formula says something is
true, then it is true," the converse was also correct. Perfection sometimes delivers bad news, and today the bad news was being carried in
Stuhlinger's hands in the form of a thousand angry formulas. He and
his fellow engineers had crunched the numbers again and again, as
if perfection could be improved upon through repetition. The answers
were always the same, and the news was always bad. Perfection was a
beautiful mistress, but she could be a real bitch.

Stuhlinger stopped for a moment to allow a golf cart full of army mucky-muck brass to pass by, then he resumed his walk. He swore in German, realizing von Braun would not be happy with the equations and results he was about to deliver, like presents from some evil twin of Santa Claus. The bad news could not have been worse for America's nescient space program: The Redstone rocket did not have enough lifting capacity to place a satellite into orbit. The final number at the bottom of the page said it all: 93.10 percent. The Redstone, as currently engineered and designed, had the capacity to come within 93.10 percent of reaching orbit—meaning, of course, that it would not reach orbit at all. It was not just off, it was *way* off. Were the Redstone to attempt an orbital launch that day, it would climb into the stratosphere, run out of energy, tilt over, and eventually splash down thousands of miles away in the Atlantic Ocean—another embarrassing failure anxiously awaiting its live television close-up.

Stuhlinger took a deep breath, then knocked on Wernher von Braun's office door.

—⋀⋀⋀—

General Bruce Medaris set the army-green receiver down on its army-green hook.

"Tell Colonel Wilkins I want to see him at once."

His secretary, Miss Biddle, stood patiently at one corner of his desk. "Where should I find him?"

"Try the golf course."

Colonel Wilkins was on the fourth green at the army's nine-hole course in Huntsville when Miss Biddle came trotting up. The appearance of the tall, skinny woman with the coke-bottle glasses always meant one thing: General Medaris wanted to see him. He tapped the ball gently and watched the trajectory skirt past the hole, missing it by several inches. Nothing like a summons from the general to throw him off his game.

The colonel excused himself from his playing partners and followed Miss Biddle.

General Bruce Medaris had two three-digit numbers in front of him. The first number was 284. The second was 305. Both numbers were measurements of specific impulse. Measured in seconds, specific impulse is a sort-of horsepower rating for rocket propellants. As with horsepower, so with specific impulse: the higher the number, the greater the power. Every fuel and oxidizer propellant combination had its own specific impulse value. Such values were calculated theoretically on paper, then adjusted downward a few points (sometimes many points) after data reduction from static-engine tests produced less than theoretical results (and they always produced less than theoretical results). In rocketry, more than any other endeavor, reality always trumped theory.

The lower number, 284, was the specific impulse for the Redstone rocket's current fuel/oxidizer propellant combination of alcohol and liquid oxygen. This was the propellant combination von Braun used in the V-2. Due to his extensive experience and knowledge with those propellants, he had chosen to utilize them in the design of his US Army rocket systems, including the Redstone. It seemed logical at the time.[1]

That number, 284, was a well-known value to von Braun and his engineers; it was one that had always served them well. However, it was also an ideal number, meaning it was the highest specific impulse one could expect from a LOX/alcohol system under the best conditions, using propellants of perfect purity and equipment of faultless function—neither of which ever happened. 284 was the ceiling, the end, the maximum, the upper limit, the pie-in-the-sky. Most important, 284 was a number no LOX/alcohol rocket could ever rise above, let alone reach.

The higher number staring at General Medaris, 305, was different. It was not theoretical. It was not pie-in-the-sky. Unlike 284, it was not a number affixed to any particular propellant combination. Rather, 305 was a number that had been vomited forth from a series of slide-rule calculations performed by Wernher von Braun's engineering team. 305 was not a number so much as it was a goal, a target, an objective, a nightmare, an impossibility. 305 seconds was the minimum specific impulse value the Redstone propellant system would have to hit if the rocket would have enough power to push a small satellite into orbit.

In other words, General Medaris had a terrific ICBM, but not the satellite booster he was seeking.

Colonel Wilkins entered, still wearing his golf shoes.

"You needed to see me, sir?"

"Yes, Colonel. I just received some disturbing news from Dr. von Braun." For a moment the general appeared lost in thought. The colonel knew better than to interrupt him.

"Turns out the Redstone, as currently designed, will never reach orbit."

"Even with the lengthened tanks and other improvements? Dr. von Braun seemed so confident at our last briefing."

"Yeah, well, the Krauts sharpened their pencils and somehow figured out that the Redstone has what it takes to get 93.1 percent of the way into orbit. Do you know what 93.1 percent gets you in the satellite business, Colonel?

"No sir."

"It gets you bupkis!" The general's face was getting red. It was not a good sign. "If we launched the Redstone today, it would keel over at a very high altitude and splash down somewhere in the Atlantic Ocean. A pile of worthless sheet metal. That's what 93.1 percent gets you. Understand?"

"Yes, sir."

"We have no idea how far along the Russian space program is, but the fact that they were able to assemble an A-bomb in no time means we ought not to underestimate them. The army needs a new propellant combination for the Redstone—a new fuel, a new oxidizer, or both. And we need 'em two days before yesterday. I'm sending you out to California—a little backwater they call Canoga Park."

"Excuse me, sir. Did you say 'California'?"

"There's a company there. North American Aviation. Got some of the best engineers in the business. Not as good as the army, of course, but good. They designed and built the Redstone booster. They know the system bass-ackwards and forwards. If anyone can solve this problem, it's them."

The general handed his subordinate a folder.

"Here's the contract. My secretary will give you a binder listing all the vendor's engineers. Their job is this: to find a new propellant combination that can give us that extra 6.9 percent of performance without making any more changes in the rocket's design. Tell them I want it two days before yesterday. I got a plane on the tarmac warming up. Dismissed."

"Sir, I probably ought to pack a few things . . ."

"You'll be back long before you soil your stupid-ass underwear, Colonel. The pilot's waiting."

"Yes, sir."

The two officers exchanged salutes, and Colonel Wilkins headed for the door.

"Colonel."

With his hand on the doorknob, Colonel Wilkins stopped and turned around.

"The country's reputation is riding on our shoulders. There's a lot at stake. Tell those white-shirts at North American I want them to put their very best man on this project. You understand? Their very best man. Whoever they choose doesn't know it yet, but he's about to become the most important person in America."

"Yes, sir."

And with that, Colonel Wilkins left the office and walked toward the far corner of the building. Long before he arrived at that corner, he could hear them—the earthy purr of four Pratt & Whitney R-1830 turbocharged radial engines, each producing 1,200 awesome, earth-shattering horsepower. R-1830 engines meant he was probably going to be riding a B-24 Liberator, one of the finest planes ever built. At a top speed of 270 mph, he would easily be back from the "backwater" in plenty of time for the annual officers-versus-enlisted men golf tournament.

The pilot gave an urgent hurry-it-up wave, and Colonel Wilkins accelerated to a trot.

—◦⟋⟍⟋⟍◦—

Many years hence, when people would ask Richard Morgan why he only bought used cars, he would say, "I'll buy a new car when the price of new cars goes down to what I paid for my Volkswagen in 1953." And he meant it. My father would hold onto that little, green VW bug with the small rear window for fifty-seven more years before one day selling it to a collector, a man who offered my father more money than the car's original purchase price. When it came to cars, I always thought my dad was crazy. Yeah, crazy like a fox.

Today I am riding in that green 1953 VW. I am a one-year-old baby.

Car seats for infants and children have not yet been invented, so my mother is holding me in her arms as my father drives me to my aunt Amy's house. By the 1950s, most of the Michael Sherman family has abandoned North Dakota and resettled in Southern California. All the Sherman kids are now grown and married and having children. My mother's older sister is now Amy Wengler and lives just a couple of miles from us in Reseda. She babysits me Monday through Friday so both my parents can go to their jobs at North American Aviation.

On this warm autumn morning, my father pulls the 1953 green VW alongside the curb in front of the Wenglers' boxy, three-bedroom stucco home. My mother hands me into my aunt's waiting arms, then gets back into the car. Since I'm only a baby, and far from sentient, I am completely unaware that my mother is about to change the world.

Of course, she is completely unaware of it as well.

—⟍⟋⟍—

Tom Meyers knew Colonel Wilkins was coming. He had been forewarned by the mucky-mucks upstairs, who had been forewarned by the mucky-mucks in Downey. It was a contract so secret no one was allowed to discuss it over the phone. A contract so important, a full-bird US Army colonel was delivering it personally.

"A representative from the army is bringing you a new contract."

"What kind of contract? A new engine?" An order for a new engine design would mean at least four more years of job security for his entire department.

"I can't discuss it on the phone, Tom. But let me make this clear: you are to put your very best man on this. The army was very insistent. Do you understand?"

"Yes, sir."

"This contract will be your department's top priority for now."

It was a mysterious call, but hardly the first such call he had received over the years. Everyone was paranoid of Soviet spies, and that paranoia was creating a great deal of anxiety and phone-a-phobia. In the aerospace industry, people were becoming wary, suspicious. Employees would walk to their cars at the end of the day and look over their shoulders to see if they were being followed. They would drive to work the next morning and take far too many glances at their rearview mirrors.

Tom Meyers hung up the phone, looked momentarily at the portrait of President Dwight D. Eisenhower that hung on his wall, then went back to work. No use wasting time while waiting for the US Army.

—◦∿∿◦—

When Mary arrived at North American that morning, she showed her ID badge, passed through the steel security door, then began the long journey to her desk. As she walked past the ranks and files of desks and files, she marveled at how many things had changed since 1946. North American had completed construction of the new Canoga Park facilities just two months before. They were ramping up for what appeared to be new ventures in ever-larger rockets and missiles. There were many new contracts, money was flowing like the Mississippi, and hundreds of new engineers were being hired.

One thing had not changed: they were all men.

And they all had engineering degrees. Mary had certainly paid her dues, making great advancements in rocket-propellant technology with both the NALAR and NAVAHO programs, but she couldn't help feeling inadequate. The days when it was possible to get a job in aerospace without a college degree were over. Thanks to time, and the GI Bill, America's pool of talented engineers was bulging. Mary was surrounded not just by men; she was surrounded by people who were much better educated than her. The situation did not bode well for job security in a company that routinely fired 5 percent of its workforce every year for no other reason than they could. As the least formally educated employee in the engineering department, she was highly vulnerable. The situation had gotten to the point that, every morning, she arrived at the lobby wondering if a pink slip would be waiting.

The morning passed uneventfully. Mary got in her usual 225 calculations on new propellant specific impulse figures before lunch. When noon arrived, she set up the card table. Joe Friedman, Carl Amenhoff, and Irving Kanarek soon joined her. She dealt out four hands.

Joe began the bidding. "Two hearts."

"Two spades," said Carl.

"Four hearts."

"Four spades," said Mary. The rest of the players passed, and she played the first card.

"I hear we're getting a visitor from White Sands."

"I heard it was Huntsville."

"No—Fort Bliss."

"It's all the same."

"Scuttlebutt says they're sending over a new contract."

Mary took the first trick. "I hope it's a new engine. All this theoretical stuff they have us on is fine, but it's always nice to build hardware."

"Just wish we could work for someone other than those government goons."

"Those government goons pay your salary."

"Ex-nay on the government-say."

Everyone turned to look. Coming down their aisle was a US Army colonel, decked out in uniform, medals, sunglasses and golf shoes. He was carrying a thick three-ring binder. When the officer arrived at the card table, he stopped.

"Excuse me. I'm looking for the Engineering Department of the Office of Research and Development."

"You found us."

"This is the engineering department?" For 3.47 seconds, no one said a word. Then the colonel spoke. "Could I ask you what on earth you are doing?"

"Playing bridge."

"They're playing; I'm winning."

"So what kind of a contract are you bringing us?"

The colonel was stoic, expressionless. Behind the sunglasses, his eyes were a mystery.

"My mission here is top secret."

"Oh, yeah—I got a question about that. What exactly is the difference between 'secret' and 'top secret'?"

"Right. How much more secret than secret is top secret?"

"If something is top secret, can it really be any more secret than just regular secret?"

"How exactly do you measure degrees of secrecy?"

"If there are different degrees of secrecy, then there must be some way of measuring those degrees."

"Exactly. Exactly! Are there units, like meters, nanoseconds, gigahertz of secrecy?"

The colonel's facial muscles did not budge. "I have an appointment with Mr. Tom Meyers."

"Up the stairs."

As the colonel made his way toward Tom's office, the three male engineers tried their best to stifle their laughter.

Mary shook her head. "Gigahertz of secrecy. You boys are bad."

—◦WW◦—

Tom returned the phone receiver to its cradle. Audra, at the receptionist desk, had called to inform him that a US Army colonel had arrived and was headed his way. Tom thanked her for the warning, then stood at the large window to watch the officer's approach. Tom cringed when the colonel made a short stop at Mary's bridge table.

Of course the guy would arrive at lunchtime. Of all the luck.

From his vantage point ten feet above the floor, Tom knew something big was coming; the three-ring binder Colonel Wilkins was carrying was at least four inches thick. Tom returned to his desk, pretending to look busy.

When the knock was heard a short while later, Tom simply said, "Come in."

Colonel Wilkins entered, removed his sunglasses, and introduced himself. The two men shook hands.

"I met a few of your engineers coming in," said the colonel. "They were playing cards."

"It's lunchtime."

The colonel checked his watch. "So it is."

"If you'd like, we can head down to the cafeteria; have our meeting after lunch."

"No, no. This is urgent. I came here as fast as the Liberator could take me."

"A B-24. Right. I flew those in Germany. I was a bombardier." This comment did not seem to impress the colonel. "Anyway, our main office in Downey mentioned you wanted to discuss a new contract."

"May I sit down?"

"Oh, yes—please."

Both men sat down and the colonel opened his binder.

"As you know, the Redstone rocket program has been a terrific success."

"Of course. We built the booster."

"Which is why I'm here. Over the last year, Dr. von Braun has been making certain modifications to the Redstone to increase its range. In order to get funding, General Medaris has been telling the bean counters these design alterations were necessary in order make the Redstone a viable ICBM. There are some technological and political hurdles that need to be overcome, but von Braun's rocket is ahead of everything else that's in the pipeline. The ICBM story, in fact, may wind up being true, but secretly the general and von Braun plan to use the Redstone to launch the world's first satellite."

"You won't get any argument from anyone around here. What can we do to help?"

The colonel cleared his throat, then opened the contract folder.

"We need to increase the Redstone's performance by at least 6.9 percent. And we need to do it without changing any more of the rocket's hardware."

Tom thought this over carefully. If all hardware was off-limits, that left only one other item.

"In other words, the only things we can change are the propellants."

"Exactly. A better fuel, a better oxidizer, or both. The magic number is 305."

"Is that our specific impulse target?"

"It is. There are some other targets that have to be reached as well."

Colonel Wilkins handed the contract file to Tom. "It's all in there: project requirements, expected benchmarks and milestones, deadlines, etcetera. Dr. von Braun and General Medaris, along with your superiors in Downey, have already affixed their signatures. The general says he wants it two days before yesterday. That was twenty-four hours ago, so now I suppose he needs it *three* days before yesterday."

"What's the rush?"

"General Medaris and Dr. von Braun both want to beat the Russians into orbit."

Tom paged through the contract requirements, puzzled. "Von Braun's people couldn't solve this problem?"

"Unfortunately, no." The colonel began paging through the four-inch binder. "I've been going over North American's list of engineers, and I notice Irving Kanarek works here. As the inventor of inhibited red fuming nitric acid—a propellant we've had great success with on the Nike—Kanarek has a blue-ribbon reputation with the army. He seems to be the most qualified candidate. I'd like you to put him in charge of this."[2]

Tom swiveled a few degrees in his chair, staring out at the engineering floor for a moment. Irving Kanarek was an intelligent and experienced chemical engineer. It would certainly be useful if the prospective project leader had prior experience in new propellant formulation. As the colonel said, Irving fit the bill. He would not be a bad choice to head up such a project.

Tom nodded. "Yes. Irving's a very good engineer."

"So may I tell the general he's our man?"

Tom did not like having to make such important personnel choices so quickly. His normal procedure involved a great deal of thought and examination. Clearly the colonel had no intention of giving him enough time for such careful analysis. In his gut Tom knew the best person to head up such a contract would need, more than anything else, experience with complex theoretical performance calculations, something that Kanarek usually farmed out to others, such as Mary. In fact, every engineer in the department sent at least some of their theoretical performance work to her. Consequently she had become the busiest person on the floor.[3]

Colonel Wilkins closed the binder. "Very well—not hearing any objections . . ."

"Hold it."

"Yes?"

"I have someone better than Kanarek."

"Better than Kanarek? Excellent. Who is it?"

"Mary Sherman Morgan."

The colonel seemed puzzled. "Mary? As in M-A-R-Y Mary?"

"That's correct."

"You have a female engineer?"

"Yes."

The colonel seemed even more puzzled. "I've studied your personnel list. I don't remember seeing anyone by that name."

Tom shifted uncomfortably in his chair. This was not going to be easy. It had been seven years since he had hired the unemployed weapons maker from Ohio. In those six years, she had proven herself to be highly skilled at calculating and predicting the performance from new rocket-propellant combinations. The other theoretical performance specialists had begun complaining that all the design engineers were using her exclusively and ignoring them. The reports he was getting from those engineers were all the same: Mary had some sort of "magic touch" when it came to rocket-propellant theory.

"She's not an engineer, per se," said Tom. "Not exactly."

"Not an engineer. Well, what is she?"

"She's an analyst. She's probably in the back section of your book, under 'Miscellaneous Personnel.'"

"Mr. Meyers, you've got nine hundred engineers on your payroll. I've read the background bio on every one of them. I really believe Kanarek is our boy."

"I disagree."

The colonel turned to the back of his binder and soon found a short description of the analyst in question. The room became very quiet as he read her bio, which took only a few seconds.

"According to this, Mrs. Morgan attended DeSales College in Ohio for two years, then dropped out to work in a weapons plant before being hired here. No doctorate, no masters; not even a bachelor's degree. No noteworthy successes or accomplishments—no published papers. Looks like the only thing we can say about this woman is that she has a high-school diploma from Ray, North Dakota—wherever that is."

"All that is true, however . . ."

"Mr. Meyers, I'm sure you know much more about engineering and rocket science than I do, but my commanding officer, General Medaris, ordered me to have you put your very best *man* on this project, and I don't think he was speaking figuratively!"

—◦◦◦—

When his desk phone rang, Don Jenkins just assumed it was his wife calling. That was the usual case, anyway. He was surprised when he heard the voice of Tom Meyers.

"Don, could you come up to my office for a minute?"

"Sure, Tom. Soon as I finish . . ."

"Right now."

Don put down the receiver, stood up, and maneuvered through the vast field of desks toward the office stairway. The word had gone out among the engineers that a "military mucky-muck" was on the prowl, and so Don went through a mental checklist of projects he might have screwed up on. The list had more than a dozen items on it by the time he turned the doorknob.

"Yes, sir?" The first thing Don noticed was that neither man looked happy.

Tom swiveled his chair in Don's direction. "A project involving theoretical performance calculations. Of all the people we have working for us, who would you choose to head that up?"

"Oh that's easy: Mary." Since she was the only woman in the building doing engineering work, Don knew a first name was all that was necessary. Still, he was unprepared for the colonel's reaction. The man stood up a little too fast, pushing his chair hard and sliding it across the office.

"A high-school diploma! A high-school diploma—for God's sakes!"

"Colonel Wilkins," said Tom, "let me enlighten you. If Mrs. Morgan had a dozen PhD's, it would not make any difference. The work she is doing here is so advanced it's not taught in any university, anywhere, at any level."

But the colonel would not let it go.

"Nine hundred engineers, all male, all with college degrees, and you're telling me the best person qualified for this job is not only the only female engineer in the company, but the only engineer without a degree. That's what you're telling me!?"

Tom kept his cool. "She's not an engineer, she's . . ."

"An analyst—yes, you said. She has no résumé, no track record. She's never accomplished anything noteworthy!"

Tom slowly stood up, eyeballing the colonel.

"Colonel, *no* one accomplishes anything noteworthy, until they do it for the first time."

For several moments, no one spoke. Then Don raised his hand.

"I still need to take lunch."

17.
WELCOME TO THE MONKEY CAGE

"I have learned to use the word "impossible" with
the greatest caution."

—WERNHER VON BRAUN

Mary was in the middle of calculating the theoretical specific impulse of several mixture ratios of pentaborane and oxygen diflouride. It was a complex exercise involving exotic propellants that would take her most of the day and required intense concentration, which is why she was not happy to be interrupted.

"Excuse me. Are you Mrs. Morgan?"

Mary swiveled her chair around to see a man wearing a "VISITOR" badge on his short-sleeve white shirt. He was of medium build and wore a tie she was sure was brand-new. His hair was jet black, and he had a US Marine Corp tattoo on his left arm. The right arm was missing past the elbow.

"Yes, I'm Mrs. Morgan."

"I'm sorry," he said. "I have to shake with my left hand."

Mary stood to shake his hand.

"My name is Nick Toby. Nice to meet you."

Kanarek was not at his desk, so Mary rolled Kanarek's chair over and motioned for Mr. Toby to sit down.

"Where did you serve, soldier?"

"I was with the 1st Parachute Battalion on Guadalcanal. Not a whole lot of jobs a man can do with only one hand. Being a sales rep I guess is one."

"What can I do for you?"

Nick handed her his business card. "I work for Lansing Chemical Company. We have a promising new chemical I'd like to give you some information on—diethylenetriamine. We're searching around for some applications."

"What is it used for now?"

"Well, nothing. That's what we're trying to do; build a market for it. Anyway, one of the chemical engineers in our lab thought it might have some possibilities as a rocket fuel."

Nick handed her a chemical product brochure.

"I was told you're the one to talk to around here—that if anyone knew whether this could be used as a rocket fuel, you would."

Mary paged through the brochure. "Well, let me look through the data, and I'll let you know. I'm kind of busy right now."

"From what I'm told, it has some attractive properties and characteristics. It's supposed to be highly energetic with most oxidizers and has a very good density: 0.9588 grams per cubic centimeter. Don't know much about rocket fuels, but I hear that's important."

"I'll get back to you."

"That would be great. Thank you. Thank so much."

They stood, shook left hands again, and then he walked away. Mary stapled the man's business card to the brochure, then threw it in the bottom drawer of her desk, where it had a lot of company. Dozens of chemical sales reps visited her every year; there was no time in her day to pay attention to every "promising new chemical" that came her way.

As Mary prepared to go back to work, she was startled by a tap on her shoulder.

"Got a second?" asked Tom.

"I'm really busy," she said, as they both sat down.

"I know you have several pet projects you're working on that are very important."

"I've got to do the data reduction on the NALAR tests, then I have the propellant calculations to do on the NAVAHO engine-propellant-reformulation project, and then . . ."

"I understand," Tom interrupted. "But they're all going to have to wait. We're pulling you off everything for a special project."

"How special?"

Tom smiled. This was the conversation he had been waiting two years to have. He laid the contract folder on Mary's desk.

"How would you like to be involved in putting America's first satellite into orbit?"

—∿∿∿—

Three years after the play *Rocket Girl* closes at Caltech, an admirer of my mother helps me create a *Wikipedia* page for her. A few months go by, and on a lark I decide to look up the page and double check its accuracy. As I read through it, I discover persons unknown have edited the article, and not in a good way. Since only a handful of people in the world know as much about Mary Sherman Morgan as I do, it puzzles me to think someone thought they could improve on it. I get on the phone and call all the sources I've been working with to see if they edited the article. Most of them are so old, they've never even heard of *Wikipedia*. It soon becomes obvious that no one who can truthfully claim to be an expert on my mother's life has edited the article, so I go back in and correct their changes. A few months go by, and my corrections have been mysteriously re-edited by anonymous individuals. I again correct the main *Wikipedia* entry. But then I notice that a "Talk" tab has been added, too. I click the link and find a separate article from a Mrs. Jack Silverman. Mrs. Silverman's article attempts to refute my mother's claim to the invention of hydyne and awards credit instead to her husband, Jack.[1]

The name Jack Silverman has come up only once during the course of my eight years of research. There was a landmark paper given at a technology conference sponsored by the American Rocket Society (ARS) in September 1955. I own a copy; it's titled "The Theoretical Specific Thrust of a Rocket Motor for the C-H-N-O-F System." This paper had nothing to do with hydyne or its invention, and the authors are listed (in this order) as: M. S. Morgan, J. Silverman, and W. T. Webber. My copy has been in my archives for more than ten years, so I am very familiar with it. During my work on the Caltech play of *Rocket Girl*, Bill Webber discussed the paper with me and mentioned that "J. Silverman" was Jack Silverman and that he had been my mother's supervisor for a short time. That was it, and I never pursued the lead further.

But now, because of Mrs. Silverman's article, pursue it I must. I call Bill Webber and set an appointment to meet with him at his house. A few days later, I drive to his home in Thousand Oaks, and he ushers me into his den.

"What's so urgent?" he asks.

"Who is Jack Silverman?"

"He was one of the supervisors at North American. Why?"

"His wife has written an article for *Wikipedia*. In it, she claims it was her husband, not my mother, who invented hydyne."

Bill's word-for-word reply: "She's full of shit."[2]

He says a few things that lead me to believe Jack Silverman, though perhaps a brilliant man, was a consummate credit-grabber. Then I remember the ARS article from 1955 and I ask about it.

"Did his name really belong on that 1955 paper as a coauthor?"

"No," said Bill. "He was just her supervisor. As such, he took advantage of his position and slapped his name on it. But your mother and I did all the work."

"So he was a credit-grabber."

"Yes."

This makes perfect sense; in all the work I've done over eight years researching and reassembling the jigsaw puzzle of my mother's life, the name Jack Silverman was never once brought up by anyone I interviewed. His name wasn't even mentioned in passing. Not one single time. Only as a coauthor of the ARS paper did I ever see his name on anything. It could be that he, like many supervisors at that time, put his name on the projects of those working under his supervision, regardless of whether or not credit was actually due to him. Despite the dearth of evidence suggesting Silverman was the inventor of hydyne (and the wealth of evidence I've accumulated over years, researching my mother's life), according to his wife's *Wikipedia* "Talk" tab, he sent "items" to NASA's archives listing himself as hydyne's inventor. If Mrs. Silverman's claim about submitting information to NASA is true, then I begin to think that the reason my mother is not as well known as she should be is that the man who had the responsibility for giving her credit for her invention chose instead to grab it for himself. And it would seem that he did it in a most unprofessional manner: by putting it into a written record and, without telling the other people involved, sending it to NASA to be permanently engraved in aerospace history marble.

I ask Bill to put the truth in writing and affix his signature to it. He readily agrees. Meanwhile, I head back home to determine how I can set the record straight about Jack Silverman.

As soon as I'm in front of my computer, I take a closer look at the identity of the wiki editors on my mother's main page. None of them appear to have the credentials to weigh in on a subject as esoteric as the invention of hydyne. One of the editors, according to *Wikipedia*'s "View History" link, is someone named Will Beback. It sounds like a whimsical alias, as in "Will Be Back." Here's the good part: a *Wikipedia* administrator has added a note that says "Will Beback" will not be back, that he or she has been "indefinitely banned from English *Wikipedia*" by the Arbitration Committee.[3]

The next day, I fire off a letter to NASA to request copies of the alleged Silverman hydyne files. Two hours later, the mailman arrives and delivers a book that I've been waiting for: Robert S. Kraemer's *Rocketdyne: Powering Humans into Space*. During my interview with former Rocketdyne engineer Bill Vietinghoff, he highly recommended it. As I casually read through it, I come to page 44 and a description of the invention of hydyne. To my surprise, Kraemer gives credit for its invention to, of all people, Irving Kanarek. In 2004, the *LA Times* had refused to publish my mother's obituary because they could find no written record of who was the actual inventor of hydyne. Now the number of "written record" claimants is starting to lengthen like a bread line in a Soviet Moscow winter.

My concern disappears, however, when I read that Kraemer lists the formula for hydyne as "75% unsymmetrical dimethylhydrazine, 25% diethylenetriamine."[4] That mixture ratio is obviously wrong; every hydyne source available lists it as 60/40, not 75/25. Plus I have had three in-person, in-depth interviews with Irving Kanarek (one on video) in which he acknowledges that hydyne was the brainchild of my mother. Even so, Kraemer's error is good news because it points the genesis of hydyne in my mother's direction; Kanarek was her immediate supervisor and even had his desk adjacent to hers. Kraemer may not be completely accurate, but at least he refutes Jack Silverman's claim.

Even so, the competing claims are disconcerting, and they become a distraction to my writing. I convince myself that my research is more than sufficient, but as I continue the work, all I can think about are those Silverman files percolating like overheated fluorine in the NASA archives. Weeks go by, then months.

The requested files from NASA never arrive.

—/\/\/—

By 5:30 in the afternoon on the first day of the contract, almost everyone had left, providing the cavernous engineering building a whisper-quiet atmosphere. Mary had spent the day reading over the contract requirements and gathering all the materials she knew she would need: chemical reference books, catalogs on commercially available chemicals and their purity levels, and a box full of reports on prior propellant studies (some of which she had written).

Fresh on her mind was a conversation she had overheard during a coffee break in the cafeteria. A group of engineers at a nearby table were in a huddle, trying hard to keep their voices from being heard. Once in a while, one of them would turn and glance at her. That would have been clue enough that Mary was the topic of conversation, but she had overheard much of the discussion.

"Better her than me."

"I wonder if they're setting her up for failure."

"Clever way to get rid of our only female engineer."

"She's an analyst, not an engineer."

"Whatever."

"I agree—I say it's a setup."

"The mucky-mucks upstairs must know it's impossible."

"*Everybody* knows it's impossible."

"Even von Braun and his best engineers couldn't solve it."

"She'll fail, then what?"

"Pink slip."

"Better her than me."

Their feelings were understandable. Though the development of liquid-propellant rocket engines was still in its infancy, certain design aspects had already been solidified into axioms. One of those axioms was that liquid rocket engines had to be designed around a specific fuel and oxidizer combination in the same way an automobile engine is designed for a specific species of petroleum. Switching one or both of the propellants after the rocket had been built was like putting diesel into a gasoline engine. What made the Redstone contract especially pernicious was that it not only required a change of propellants that would work with the existing design and hardware, but the change had to yield a major improvement in performance. From an engineering standpoint, it was an exceptionally egregious demand—no rocket system had ever been called upon to do this. The contract simply had no precedent. In the minds of the von Braun–worshipping engineers, if Wernher couldn't solve a rocket-engineering problem, no one could. In all the world nobody knew more about the subject than he. Now here was a farm girl from North Dakota, with nothing more on her wall than a high-school diploma, being told she had to succeed where experience, education, and genius had all failed.

Mary had watched the huddle out of the periphery of her vision. She could see the pity in the eyes of her coworkers. But the pity was mixed with relief—relief that they had dodged the bullet—that none of them had been chosen for an assignment that could well be described as engineering suicide. Under Kindelberger's "5 percent rule," anyone who failed at such an important task would almost certainly get pink-slipped.

After lunch, her immediate supervisor, Irving Kanarek, had approached her desk with some advice.

"As you know, the Redstone uses liquid oxygen. There are several oxidizers out there we could use much more powerful than LOX. I recommend you focus on keeping the existing fuel but switching out to a different oxidizer. I talked to Tom, and he agrees with me on this."

She had considered his suggestion all afternoon, making a list of oxidizers that might provide a specific impulse of at least 305 with alcohol. There were only a few, and all of them had problems of the worst kind. She even considered FLOX (fluorine/oxygen mixture), but the plumbing and turbopumps of the Redstone engine were not designed for an ultra-corrosive, ultra-reactive chemical like fluorine. Upon firing, she estimated the engine would run for approximately 250 milliseconds before exploding. Fluorine was out, even in a diluted mixture form.

And so, without telling anyone, she quietly turned her attention to the fuel side. The odds of finding a superior fuel were much higher, given that the natural world of atoms and molecules allowed for the creation of only a handful of good oxidizers but hundreds of fuels.

Mary took out a pad of paper. At the top she wrote "Properties and Characteristics." Beneath that she wrote a column of numbers; 1 through 10. To the right of number 1 she wrote "Commercially Available."

There are ten properties and characteristics our new propellants must have, she thought. *First and foremost they must be commercially available. The world's greatest fuel and oxidizer do us no good if we can't buy them someplace.*

Next to the number 2 she wrote "Physical Data."

We have to know its physical data. We can't work with a chemical we know nothing about. That's obvious.

Number 3: "Vapor Pressure."

Our fuel has to have a low vapor pressure; it must be a liquid at ambient temperatures and pressures. The fuel side of the system is not designed to work with a cryogenic liquid like the oxidizer side. The fuel valves would freeze right up.

"Freeze right up—just like a North Dakota prairie winter."

Frightened, Mary stood up. "Who said that!?"

She looked around the engineering floor, flat as the side of a barn. There was a janitor sweeping that floor, but he was two hundred feet away. Other than that, she was alone. She felt a cold breeze from somewhere, and then the smooth concrete floor of the building disappeared, and in its place was the cold prairie grass of North Dakota. She was wearing her knee-high dress; her legs and feet were bare. The arctic wind made goose bumps rise on her arms and legs.

Clarence was standing nearby, holding a switch.

"You should never have left the farm, Mary. Mother and father are angry."

He came at her fast, grabbing her arm with his free hand while swinging the switch at her legs with the other. His swipes were swift and violent and many. Liquids poured from her body; tears from her eyes, blood from her thighs.

"You should never have left the farm, Mary! Mother and father are angry!"

"Please stop!"

"Mother and father are angry!"

"Please stop!"

"You should never have left the farm! You should never have left the farm!"

I should never have left the farm.

A low-frequency hum intruded, and slowly Mary lifted her head from her folded arms and rejoined the concrete world. Mary opened her eyes to see a second janitor maneuvering a floor polisher nearby. The man was looking at her, a strange expression on his face.

"Are you okay?" he asked.

Mary nodded. *I must have fallen asleep.*

"Hey!" she heard someone shout. Mary turned to face the sound. It was Richard, a wide smile on his face, about fifty feet away and walking toward her. Between the two of them they had only one car; he was here on schedule to pick her up.

"Ready to go?"

Mary shook her head and sat back down as he arrived at her desk. "No. I have to finish my list."

Richard looked over her shoulder. "Properties and characteristics. Working on a new propellant combination?"

"Yes, but I don't know what it is yet."

"You don't know what it is? Isn't that the opposite of the way it normally works?"

"Yes. It's sort of a reverse-engineering project."

"Need some help?"

"No. Sit down and don't interrupt."

Richard sat down in Kanarek's chair. "Okay. But we have to pick up George pretty soon."[5]

Mary nodded. She had seven more properties and characteristics that needed her attention, and she wanted to get them down on paper while they were fresh in her mind.

Availability. Physical data. Low vapor pressure. Let's see . . .

Number 4: "Mixture Ratio."

The mixture ratio is already set by the Redstone's plumbing and design—it can't be changed. The engine is designed to use about 1.75 pounds of oxidizer for every one pound of fuel. Our new propellant must optimize at something close to that or we won't get the maximum energy from combustion, and we won't reach the magic number of 305.

Richard found Kanarek's deck of playing cards and started laying them out for a game of solitaire.

Number 5: "Stability."

If at all possible we want to avoid unstable chemicals due to all the problems with handling and storage.

Richard put a red four over a black five.

Number 6: "Controllable toxicity"—*for the same reasons.*

The black jack went over the red queen.

Number 7: "High Heat of Combustion."

The hotter the flame, the more energy we gain.

Richard was stuck. He started going through the deck. "If this propellant is going to be used in a regeneratively cooled engine you'll want to list 'good heat-transfer properties.'"

Number 8: "Good Heat Transfer Properties."

Numbers 9 and 10: "Low Molecular Weight, and a High Ratio of Reactive Atoms"—*such as hydrogen, in the molecule. That's it; that's the list that determines our work.*

Mary put a heavy chemical reference book atop the list to make sure it didn't grow legs and walk off during the night, then grabbed her purse.

"Okay. Let's go get George."

"I'm not done with my game."

Five minutes later, as the 1953 green VW with the small rear window was negotiating the streets of the San Fernando Valley, Mary could not shake the feeling that she had forgotten something on her list.

Properties and characteristics. There's something I'm missing.

Mary dragged herself and two cups of coffee to her desk. It had been a long night, and she hadn't slept much. Thousands of specific impulse and propellant performance formulas had been racing through her thoughts. The more she considered the problem, the more she had come to agree with the lunch gossipers; everything about the contract requirement seemed impossible. Alcohol was a damn fine fuel, and every pipe and valve and injector in the Redstone was designed specifically for it. On the way home from work, she asked her heat-transfer-specialist husband's advice.

"What if we simply changed the alcohol from a seventy-five percent mixture with water to ninety-five percent, or one-hundred percent. Anything wrong with that?"

Richard had considered the suggestion carefully, then rejected it.

"That would raise the temperature of combustion substantially, but that's not a big problem. I'm sure the A-7 engine could handle it. The real problem is you wouldn't get a three-oh-five isp doing that. Even pure alcohol is under three hundred, I believe."

She checked his answer, and it turned out to be correct.

If the solution to the Redstone power problem were that simple, the Von Braun boys wouldn't need us.

She set one coffee cup down and started drinking the second. It was the morning of Day 2 of what Mary had started calling MUPP: the "Mysterious Unknown Propellant Project."

Tom arrived at her desk just as she sat down.

"I've recruited two engineers to assist you. They're getting some paperwork squared away, then I'll bring them over."

"Good," said Mary, "because this is going to be a massive amount of work."

"I know."

"In fact, I could probably use eight or ten."

"Ya got two."

He was about to leave, then added, "Did Irving tell you what we talked about? That we both feel you should be focusing on the oxidizer side?"

"Yes, he told me."

Then he left.

Immediately after deciding to focus on the fuel side, Mary had chosen not to tell her supervisors that she intended to go down a different path. One thing she had learned since working side by side with nine hundred men was this: Never tell them they're wrong until you can prove it. And she was far from proving it. From her experience with engineers, she knew they hated being told they were wrong, but didn't seem to mind being proven wrong. In the final analysis, all engineers and scientists really cared about was the discovery of truth; you just had to be ready to prove your position.

An hour later, Tom returned with two young men in tow. One had pale skin and wore the de rigueur engineer's crew cut. The other was a Japanese man who looked several years older.

"Mary, I'd like you to meet Bill Webber and Toru Shimizu. Gentlemen, North American's finest propellant analyst, Mary Morgan."

The three shook hands. Mary noticed their badges looked brand-new. Had they been recruited from another division or another department of the company?

Tom pointed to two adjoining desks. "Those will be yours. I leave you in capable hands."

The two young men looked lost. To Mary, it seemed that they had the air of a couple of high-school freshmen on the first day of class.

"Why don't you two pull your chairs over and we'll have a meeting."

As they did that, Mary asked, "So which department did they move you over from?"

Bill and Toru looked at each other, clearly unsure of what to say. Bill spoke first.

"Well, I just got my master's degree in chemical engineering from Caltech. I was at a recruiting fair and North American hired me."

Mary was surprised. "You're right out of school?"

Bill nodded. Mary turned to Toru, hoping for more.

"And what about you?"

"I just received my MS in chemical engineering from UCLA."

There was a long pause as everyone worried over a different problem, then Toru added:

"I would have graduated sooner, but I spent four years in Manzanar."

"Manzanar. You were in a Japanese internment camp."

Toru nodded. Mary rubbed her temples—a headache was coming on.

"So for my assistants I have been given a couple of virgin newbies."

At first Mary felt indignant about being given two assistants with no rocket or propellant experience whatsoever. That lasted about five seconds, at which point she changed her mind and decided it would be advantageous to work with fresh meat—a couple of guys who carried none of the baggage of preconceived notions of what was possible. For this project she would need open minds. Besides, even though she had far more experience, these two young men were far better educated. She would find their strengths and put them to good use.

Mary handed each man a file full of graph paper, on which was a myriad of formulas and math calculations.

"In the propellant-design field we work a great deal with simultaneous nonlinear equations. Have either of you ever done any of those?"

Toru spoke first. "Well, there's no straightforward method for solving simultaneous nonlinear equations."

"That's correct," said Mary.

Bill asked, "So what's your approach?"

Mary smiled. They may have been new recruits, but she could tell they had potential.

"We do a lot of guessing. We make an intelligent guess at what a particular combustion temperature will be based on past experience and measurements, then calculate what the composition would be without the dissociation. Then we calculate how much dissociated species would be present by repeated approximations. Once that's done, we calculate the enthalpy balance to determine whether our temperature guess was too high or too low. Then . . ."

"Then you make a better guess of the temperature and continue to repeat the process until everything balances out," said Bill.

"Exactly."

Toru raised his hand. "But all that work isn't going to yield much data."

"Correct," said Mary. "The bad news is that 'all that work,' as you say, only yields a single data point on a very long curve for a single propellant combination using a single mixture ratio at a single chamber pressure."

"The permutations must be endless," said Toru.

Mary nodded. "They're infinite. And we don't have time for infinite."

"So what you're saying," said Bill, "is that our country's success or failure in getting into space is highly dependent on our ability to make good guesses."

"That's pretty much it." Mary pulled a Kent from its package and lit it with a match. "The good news is that the Russians have the very same problem."

Bill shook his head. "You could get a damn monkey to make a halfway-decent guess."

Mary took a long drag on the cigarette and blew the smoke out long and slow.

"Welcome to the monkey cage."

THE MYSTERIOUS UNKNOWN PROPELLANT PROJECT

> "There is no such thing as consensus science. If it's consensus, it isn't science. If it's science, it isn't consensus."
>
> —MICHAEL CRICHTON

The next morning as Mary was making breakfast, she spread some honey on a piece of toast. The image of the golden honey triggered a long-ago memory, a flashback: Mrs. Bowman's chemistry experiment in the one-room schoolhouse in Ray. Water, vegetable oil, honey. It was an experiment about buoyancy and . . .

"Density! How could I forget density?"

She tore off a section of the morning newspaper and wrote "HIGH DENSITY" in large block letters, then placed the note in her purse. The sizes of the Redstone propellant tanks were fixed, so whatever propellants they used would have to be dense enough to fit. Therefore, density of the final propellant was crucial.

An hour later, Mary was showing her badge to the security guard at North American, passing through the steel door, and negotiating the desk maze to her station. She had arrived an hour early so she could get a few things done before Bill and Toru showed up. But when she reached her desk, she was surprised to find both men already there. On all three desks were mounted large cardboard posters with fuel and oxidizer names and performance data.

"How long have you two been here?"

"A couple hours."

"We wanted to get a head start."

"We put together some potential propellant combinations."

"If you have a moment we've prepared a little presentation."

"You guys came in early to make a prospect list. You've impressed me."

They smiled, proud of themselves—until Mary shook her head.

"You've got all sorts of oxidizers on here. I told you: we're focusing on the fuel side of the system." She grabbed a black marker and pulled off its cap.

"I'll save you boys some work."

Using the marker, she began crossing chemicals off the list.

"Let's see, you've got alcohol with fluorine. I know what you're thinking; the Redstone is already alcohol-compliant, so why not lean toward it. But the oxidizer side of the system will not withstand even a slight amount of fluorine. So we'll cross out fluorine and all its derivatives like FLOX; already considered and dismissed. What else ya got here? Hydrazine—no. Monomethyl hydrazine—no. Aniline with ozone; that would give us a good isp but ozone is too unstable—that's a no. Propane with LOX—no. JP-4 with LOX—wrong mixture ratio. Same problem with all the kerosene pairs you have here—no, no, no. Hydroxyl-terminated polybutadiene—my God, how did you even know about that one? Mixed with fluorine—already a no. Matched with nitrous oxide—isp will be too low. Ethylene with LOX—not bad, we'll hold on to that one. Ammonia with LOX—no. B2H6 with hydrazine as an oxidizer? Innovative, but hydrazine is out on both sides. Aniline with RFNA—no. Hydrogen . . . hydrogen!? What have you two been drinking? Methane with LOX—that might take us somewhere. Lithium with fluorine!? My god, you boys are dangerous. RP-1 with nitrous oxide—no. Turpentine with nitric acid—no. And finally, nitrogen tetroxide and pentaborane with LOX—no and no."

The posters were now covered in black ink *X*'s, with every chemical formula combination crossed out save for two. Bill and Toru looked disappointed.

"We worked hard on that."

"Why were you so quick to dismiss hydrogen? The specific impulse with oxygen is terrific."

"And what's so bad about hydrazine? My reference book says . . ."

"Your reference book has never built a rocket. Sit down."

Both men sat down.

"First of all, let's get hydrogen out of the way. We absolutely cannot use it. Anyone want to guess why?"

Bill spoke first. "It's cryogenic."

"So I take it the fuel side probably would not accept a cryogenic liquid," said Toru.

"You're both right, and you're both wrong."

Mary pulled the "properties and characteristics" list from beneath the chemical reference book, added "density" at the bottom, then held it out so both men could read it.

"This is the list of properties and characteristics our new fuel must have. What's the first item on the list?"

"Commercially Available."

Mary taped the list to a file cabinet. "Let's say we got past the cryogenic problem and found a way to use hydrogen. Where would we buy it?"

Both men looked at each other, hoping the other would have the answer.

"Well, I'm sure there must be some place . . ."

"No. Liquid hydrogen is not available in large quantities in the United States, or in any country in the world we can import from. It's not commercially available anywhere. Someday it probably will be, but if you need more than a few liters right now, you're out of luck."

She let that sink in, then continued.

"Now let's move on to your other question: What's so bad about hydrazine? The A-7 engine is regeneratively cooled. Do you boys know what that means?"

"Not exactly."

"It means the fuel is first circulated around the rocket engine before being pumped into the injector and burned. This helps keep the rocket engine at a stable temperature. Or to put it another way, it keeps the engine from overheating and blowing up."

"In other words, same principle as water cooling in a piston engine."

"Exactly, except cars don't run on water. Here we're using the actual fuel to do the cooling."

"Right."

"So one of the properties and characteristics of hydrazine is that it is a very poor coolant. It does not have good heat-transfer qualities."

"So if we used it the engine might overheat."

"I'm afraid so. We tried using it in the NAVAHO engine, with negative results."

"Define 'negative.'"

"It blew up."

"Ah."

Mary snatched the list from the file cabinet and handed it to Toru. I want you guys to make twenty copies of this list and study it carefully. After lunch we'll do what you tried so nobly to do this morning: we will make our prospect list.

"Why do we need twenty copies?"

"I'm going to have you pass the list around to some of the other engineers and supervisors to make sure I haven't left anything off." Mary handed them each a list of the individuals she wanted the properties and characteristics distributed to.

"Mrs. Morgan," said Toru. "Do we really have to get other engineers involved? Bill and me, we talked it over and we're pretty confident the three of us can achieve the contract requirements."

Mary looked at Bill, who seemed to agree.

"Look, guys, when you're in school it's all lonely work. You study alone, you do your homework alone, and when the exam comes around you have to complete it alone. But once you're in the workforce the rules change one-eighty. Now we work together as a team, and everyone in this building is part of our team, even if they aren't specifically assigned to us—understand? Now get that list copied off, and make sure everyone on the second list gets a copy. One more thing."

She handed them a book. It was titled *Rocket Propulsion Elements*.

"This is an advance copy of a book that's about to be published. It was written by one of our engineers, George Sutton. Go to the chapter on specific impulse. Don't come back here till you understand what it is and how to calculate it. Now scoot!"

As Bill and Toru headed off for the mimeograph station, Mary took the stairs to Tom Meyers's office.

—WWW—

At about the same moment Mary was climbing the stairs to have a word with her boss, Wernher von Braun, his wife, and 102 fellow German scientists and their families were being sworn in as naturalized US citizens. As part of the ceremony, von Braun was invited to give a speech.

"This is the happiest and most significant day in my life," he said. "I must say we all became American citizens in our hearts long ago. I have never regretted my decision to come to this country."[1]

The ceremony was held at Huntsville High School's auditorium, and there was not an empty seat to be had.

Afterward, von Braun jumped right back into his daily grind: crisscrossing the country preaching the gospel of space travel, making speeches, appearing as a guest on the Walt Disney TV show, and working with other engineers to resolve all the technical barriers to putting a satellite into orbit.

Of those technical barriers, none loomed as large or as urgent as the main-stage propellant problem. In fact, the propellant problem had become the only problem—the bottleneck that was keeping him and his rocket from reaching orbital-space glory. For this reason, von Braun decided to travel to California and pay a visit to the executives at North American Aviation.

—WWW—

Josef Stalin had been dead for a year. His successor, Nikita Khrushchev, was still working on solidifying his place in the Soviet government.

Unable to afford an arms race with the United States, Khrushchev was gambling that Korolev's missile program would make moot America's dominance in bombers and other heavy aircraft. Ironically, history would show that Russia's inability to compete with the United States in financial resources would cause them to stumble into a much superior technology. Korolev would not disappoint his premier—Khrushchev's gamble would pay off in ways more spectacular than anyone could imagine.[2]

As Wernher von Braun was flying to California to have an urgent meeting with the executives at North American Aviation, Nikita Khrushchev and a small group of Soviet leaders were traveling to Tyuratam to meet with Chief Designer. They wanted to see for themselves what their rubles were buying at his rocket factory.

—∿∿—

When Bill and Toru returned, Mary was not at her desk. The first thing they noticed was that a large chalkboard on wheels had been brought in and was parked in front of Toru's desk. On it were chalked four numbers: 305, 1.75, 155, and 0.8580. The second thing they noticed was that someone had crossed out the last two propellant combinations on their posters.

"Do you feel as unappreciated as I do?"

"Yes."

At that moment, Mary returned, carrying a box full of files.

"You decided you didn't like any of our choices."

"I like all of them, but Mr. 305 and Mrs. 1.75 dislike them immensely." Mary set the box down on her desk, then stood next to the chalkboard. "So are you both comfortable with specific impulse?"

Both men nodded. "We get it."

"Good. So here's our job: Von Braun needs a propellant combination that will yield a specific impulse of at least 305 seconds. The Redstone hardware is designed to mix the propellants in a ratio of 1.75, meaning for every 1.75 pounds of oxidizer that flows into the combustion chamber, one pound of fuel flows in to mix with it. Whatever propellant combination we end up with, it must yield an isp of at least 305 when mixed at a 1.75 ratio. Understand?"

Both men nodded.

"Which is why I crossed out all of your suggestions. None of them will give us an isp of 305 if mixed at a 1.75 ratio. They all have the ability to give us impressive specific impulse results, but at vastly different mixture ratios."

Toru raised his hand. "What is the third number?"

"The third number is burn time. The Redstone was originally designed with a first-stage burn time of 110 seconds. But that was in its original genesis as an ICBM. To get into orbit, von Braun has calculated that he will need to extend the first-stage burn to 155 seconds."

"Why is that important to us?"

"Good question, which brings us to the final number: 0.858. The units here are grams per cubic centimeter."

"Density?"

"Exactly. Whatever fuel we decide to replace the alcohol with must have a density of at least 0.858. Otherwise we won't be able to put enough into the tank for a 155-second burn."

"You're saying it has to be even denser than the alcohol, which is around 0.80."

"0.7893 to be exact."

"So where do we start?"

"At the beginning. We need to make a prospect list—a list of all chemicals that yield an isp greater than 305 when mixed with liquid oxygen. It will be a process of elimination. We will strike fuels from the list one by one until we find the fuel that conforms to all four of those numbers."

Both men were quiet for a moment, then Bill spoke.

"Well, uh, when Toru and I were handing that list out, like you asked us, we ran into Tom Meyers."

"And John Tormey," Toru added.

"I guess Tormey is Tom's supervisor, right?"

"Yes," said Mary. "So?"

"Well, they said we're supposed to be looking for a better oxidizer, not a better fuel."

"What did you tell them?"

Toru shifted in his seat. "We didn't really tell them anything."

"We weren't sure what to say. But if two managers, both with lots of authority and very expensive suits, tell us to look for alternative oxidizers, rather than fuels, shouldn't we be doing that?"

"No," said Mary. "Now let's get to work."

—◦〜〜〜◦—

One day I get a call from Delores Bing, a member of the performing-arts faculty at Caltech. She is calling with good news: the Caltech Theater Department has hired a replacement for the recently retired Shirley Marneus. Caltech once again has a theatre arts director. She tells me the reason the hiring process took several months was that they went through a large number of prospective directors.

"We're still committed to doing your play," she says. "And we think you'll like who we've chosen to do it. He has a lot of film and TV credits."

"What's his name?" I ask.

"Brian Brophy."

I'm a member of the Writers Guild of America, so I can drop Hollywood names with the best of them, but Brian Brophy doesn't ring any bells.

"He'll be getting in contact with you soon," she promises.

Delores and I say our good-byes, and I hang up. Right away I search for Brian Brophy on IMDb and discover that he played the role of Commander Maddox in the *Star Trek Next Generation* episode "Measure of a Man." It's one of the most well-known and admired episodes of the whole series.

A couple of weeks go by, then Brian calls me. We set up a meeting on the campus of Caltech. We meet, we greet, we talk shop, we "do lunch" at the Caltech Athenaeum. We have the buffet, and during our meal I find out he taught theater at Cal State Los Angeles at the same time my daughter Averie was a theater major there. So the conversation proceeds in the next most logical direction.

"My daughter went to school there at the same time."

"Really. What's her name?"

"Averie Morgan. She has bright red hair—you wouldn't be able to miss her. Did you ever meet her?"

Brian sits bolt upright and stops chewing his food.

"Averie Morgan is your daughter?"

"Yes. Did you know her?"

"She was in one of my classes."

"Did she give you any trouble?"

"Hmm. I'd rather not talk about that."

And that was the end of that conversation.

When *Rocket Girl* opened at Ramo Auditorium on November 7, 2008, I could not have been happier with the results. Brian turned out to be a terrific director, the actors were excellent, and great attention was paid to the stage set—a section of the mid-1950s engineering floor at North American Aviation. On closing night, when my eighty-two-year-old father came to see the play, I asked him about the set.

"The only thing you have to do to make that set more true to life," he said, in his elderly, raspy voice, "is push the desks closer together."

—⋀⋀⋀—

"Unsymmetrical dimethylhydrazine," said Bill, as he and Toru rolled their chairs over to Mary's desk.

"It will give us an isp greater than 315 and the mixture ratio is almost identical to alcohol."

They handed Mary a paper with their calculations.

"Unsymmetrical Dimethylhydrazine," she said. "One of my favorite propellants. More frequently referred to as UDMH."

"You can thank the Soviets—they invented it."

"*Spasiba*," said Mary.

"Huh?"

"*Spasiba*. It means 'thank you' in Russian."

Bill rolled his eyes. "I'm not going to ask how you know that."

Toru pointed to their calculations. "So, UDMH. Really powerful stuff. Is there any reason we can't use it?"

Mary handed the papers back. "Why did you even bring this to me? Its density is only 0.7914. It's way off."

"Yes, but we called Huntsville, and one of the Redstone engineers says the fuel tank is rated to hold a pressure of 20 psi."

"So you're thinking if we load the tank with UDMH, and pressurize it before launch . . ."

"Its density will increase slightly. It's so close in every way except density."

"Interesting idea. The question is, can we bring the density of UDMH up to spec through low pressurization. Crunch the numbers and tell me what you get."

—◦◦—

Sergei Korolev had built the largest, most powerful rocket in history. Dubbed the R-7, it was designed as both a Russian ICBM and a satellite launch vehicle. On a cold Kazakh Desert morning in May, the chief designer and his crew were preparing to push the button for the R-7's maiden flight. The test was planned strictly as a ballistic flight—the equipment needed for an orbital mission had not yet been installed.

One of Korolev's deputies, Leonid Voskresenskiy, was the supervisor in charge of loading the propellants. While his team was engaged in loading the liquid oxygen, Leonid noticed a leak in one of the valves. As his men puzzled over how to resolve the problem, Leonid took the initiative. Standing next to the rocket, he unzipped his pants and urinated all over the valve. The –297-degree temperature of the oxygen immediately froze the urine and plugged the leak.[3]

"There!" he said. "No more leak."

The rocket was cleared for takeoff.

—∧∧∧—

Bill and Toru wheeled their chairs to Mary's desk.

"What do you have?"

"UDMH won't work. We wouldn't be able to increase the density more than about one percent."

"I know," said Mary.

"You know? If you knew, why did you put us through the exercise?"

"Because I could have been wrong. What else are you working on?"

Toru handed her more calculations. "Ethylene diamine."

"What about it?"

"We think this has a lot of potential."

"It complies better than anything we've found: isp, mixture ratio, and density."

Mary shook her head. "Those three factors are not our only consideration. Remember the original 'properties and characteristics' list I gave you. Our propellant has to comply with most of the things on that list. The only one we might be able to bend on is toxicity; other than that, you have to follow the list."

Toru turned to Bill. "Is there something ethylene diamine doesn't comply with?"

Bill removed the list from his pocket and looked it over. "Oh crap."

"You figure it out?" she asked.

Bill nodded. "Its boiling point is too low. We'd have to put it under a lot of pressure in order to keep it in a liquid state."

"Which we can't do. Keep at it, guys."

Bill and Toru returned to their desks.

—∧∧∧—

"You want us to do *what*!?" asked Wernher.

"I want you to fill the top stage of the Jupiter C with sand," said General Medaris.

"Why would you want us to do that?" asked the German rocketeer.

"It's not me," said the general. "It's the Pentagon. They're so damned paranoid that one of your test rockets may 'accidently' go into orbit that they have ordered us to intentionally weigh it down. They're putting all their eggs in the Vanguard basket. It's a big political football."

"Accidently go into orbit." Wernher smiled. "Well that would be the beauty of it."

Both men laughed.

"Look, Wernher, I understand this is utterly absurd. But I'm a military man, and what I do more than anything else is obey orders. I'll have a truck full of beach sand waiting at the launch site. Just make sure the technicians load it into the rocket."

"You Americans are a strange group," said Wernher.

"Wait till you live here a while." With that, the general turned on his heels and left the office.

From his window von Braun could see the new version of the Redstone resting horizontally on a railroad car. It had three stages now, and so they had given it a new name: the Jupiter C. The rocket was packed and ready for shipment to Cape Canaveral, where it would be undergoing a routine flight test. But nothing was ever truly routine in the life of Wernher von Braun. The new version of the Redstone was America's best option for getting into orbit, and now his adopted country was once again putting him and his hardware on the shelf.

Wernher took the number-2 pencil he was holding and broke it in half.

—∿∿—

"What about diethylamine."

Bill and Toru had been investigating every compound in the four-inch-thick chemical reference book, looking for the Mysterious Unknown Propellant. Today they were certain they had found one that would satisfy the contract requirements. All they had to do was convince Mary.

"Diethylamine," Mary repeated. "I'm not familiar with it." Mary reached out her hand, and Bill passed her a list of the chemical's properties and characteristics.

"DOW Chemical makes it in large quantities—no problem with availability," said Bill.

"Boiling point is above 131 degrees," added Toru.

"We get an isp of more than 320 with LOX, and the mixture ratio is in the ballpark."

"And the density?" asked Mary, with more than a hint of doubt.

"0.70, but listen: we're thinking the density becomes less crucial if we can get a far higher isp, which in this case we can."

"Exactly. What does it matter if we run out of fuel five seconds early as long as we get enough performance out of the rocket to get us as high and as fast as we need?"

"Good thinking," said Mary. "I give you an 'A' for innovation."

Mary stood up and stepped over to the chalkboard. "Let's crunch the numbers."

The chalkboard was filled with chemical formulas. As Mary began erasing them, Bill could not help but notice many of the formulas involved unsymmetrical dimethylhydrazine, a fuel they had considered and discarded four weeks before.

What is she working on?

●─⋁⋁⋁─●

Leonid Voskresenskiy produced a bottle of cognac and passed it around to the leaders of the Soviet R-7 rocket team. After five consecutive failures, the latest R-7 rocket had flown almost flawlessly, hitting its target like a bull's-eye near the Pacific Ocean. Chief Designer Sergei Korolev was in an especially upbeat mood—it was nothing less than the best day of his life.

As the alcohol coursed through their bloodstreams, and as its concentration became ever higher, the men dared to think the unthinkable: they would launch a satellite into orbit, and do it with or without the blessing of Moscow.[4]

Someone put on some music, and the men began to dance.

•–⋀⋀⋀–•

Bill and Toru returned from lunch to find Mary standing in front of the chalkboard, staring at it with a blank expression. The chemical formula for UDMH was written there, along with a list of its properties and characteristics.

"We're back to UDMH," said Toru, a trace of exasperation in his voice. "We've considered and discarded that propellant half a dozen times."

As often happened, Mary was lost in a world of intense thought. Bill and Toru had learned not to disturb her during these focus moments, so they quietly sat in their chairs and waited. After a few minutes, Mary returned to cognizance of her surroundings.

"Remember that first idea we were kicking around—the one about how UDMH would be the perfect propellant if only we could make it denser?"

"You said we couldn't do that."

"Not with pressure. But there's another way."

"We're listening," said Bill.

"What if we took a second fuel that had a high density and mixed it with the UDMH. You'd get the benefit of both propellants—the higher performance with the UDMH, a higher density thanks to your mystery additive."

"The key word in that sentence is *mystery*," said Bill.

"Is that what you're suggesting?" asked Toru.

"Yes. That's exactly what I'm suggesting. We've been at this for two months now, and we've got nothing. I think we all know there is no off-the-shelf compound that's going to give us what we need. The next logical choice is a cocktail."

"So what do we do now?"

"Now we make another prospect list: fuels that are miscible with UDMH, but have a high density."

Both men groaned. The moment had a very starting-all-over feel to it.

"What if this doesn't work?"

"It better, if we want to keep our jobs."

"Okay," said Toru. "I suppose you want the list by tomorrow, right?"

Mary shook her head. "You've got one hour."

As the men returned to their desks, Mary felt a stab of nausea. It quickly worsened, and so she ran to the rest room, making it to the toilet just in time. A moment later, she was spitting out the scrambled eggs and coffee from breakfast.

If I didn't know better, I'd swear that felt like morning sickness.

—⋀⋀⋀—

Collier's was having a heyday with subscriptions and newsstand copies of its magazine. Ever since it had decided to have Wernher von Braun write space-related articles for it, the publication had become the envy of the industry—copies were "rocketing" off the shelves. The latest article involved von Braun's imaginative ideas on what it would take to send men from Earth to Mars and back. It was filled with colorful artwork and, like all the previous articles, was generating a legion of fans for the German expatriate.

Von Braun had just received his first copy of the latest edition. He read through it carefully, noting places he should have written better, or more clearly. On this particular morning, he was seated in a chair on the observation deck of the Canaveral lighthouse, which had the best view of the launch area. The creation of NASA was still several years away, and Cape Canaveral was little more than swampland whose chief denizens were mosquitos and alligators. His two-way radio came alive, and a voice said the countdown had reached minus ten seconds. He finished the paragraph he was in the middle of, then looked up just in time to see a Redstone test rocket fire and launch high into the clouds.

"*Das ist gut,*" he said, then he returned to the article.

—⋀⋀⋀—

"We have a prospect list," said Bill.

"I made one, too," said Mary. "Tell me what you have."

Toru handed her a paper. "The main problem is there's not enough data about miscibility in some of these."

Mary nodded. "Right. I found the same problem."

"The best options we can see are hydrazine, pyrrole—C_4H_5N, and furan—C_4H_4O."

"You left out aniline. And what about pyrrolidine—C_4H_9N?"

Bill looked at his copy. "Yeah. Well, you only gave us an hour."

Mary began to feel queasy again. "Look, why don't we take a lunch break."

"We took lunch two hours ago."

"Well *I'll* take lunch, then." Mary jumped out of her chair and ran toward the restroom.

The next morning, Bill and Toru arrived at their desks to find Mary standing in front of the chalkboard, once again staring at the chemical formula for unsymmetrical dimethylhydrazine. The "mystery additive" prospect list was written in a column on the right side of the board. Every one of chemicals on the list was crossed out.

"Do you feel as unappreciated as I do?"

"Yes."

The sound of Mary whispering to herself was soft, but distinct. They approached with caution.

"Nine-five-eight-eight," she whispered.

"What was that?" Bill and Toru stepped closer.

Both men and the entire world were tuned out as Mary continued to stare at the chalkboard.

"Nine-five-eight-eight."

"I'm guessing that's not a batting average."

Mary turned and looked at them. "Nine-five-eight-eight." She closed her eyes, deep in thought. "I think that pencils out."

"You want to let us in on what nine-five-eight-eight is?"

"I think we found it. Now all we have to do is find it."

"Huh?"

Mary pulled open one of her desk drawers and started frantically rummaging through it with one hand, pointing with the other. "Bill—look through that file cabinet. Toru—take that one."

"It would help if we knew what we were looking for."

"A brochure!" Mary removed one of the desk drawers and dumped its contents onto the floor. "A chemical brochure!"

"From what company?"

"I can't remember!"

"What chemical?"

"I can't remember! But it's in the amine family. For some reason or other I remember its density: 0.9588."

"You can't remember the chemical, but you remember its density!?"

"Anything else you can tell us?"

"The brochure has a business card stapled to the front of it." Mary dumped out another desk drawer.

As Bill and Toru began searching the file cabinets, one thing became obvious very quickly: almost all the chemical brochures Mary had filed away had business cards stapled to them.

"So what's so special about this amine compound?"

"Listen! A density of 0.9588. And it's an organic compound, which means it's probably miscible with unsymmetrical dimethylhydrazine."

"So we're looking for our cocktail partner."

"Exactly. And we *have* to find it!"

Toru held up a brochure. "Is this it?"

Mary leaned in for a close look. "No." She was about to go through another drawer when she noticed something under his foot.

"That's it!" Mary bent over and pulled the brochure from underneath his shoe. She read the name off the cover. "Diethylenetriamine."

"Diethylenetriamine. Never heard of it."

"*No* one's ever heard of it," she said, walking over piles of dumped brochures to get to her phone.

•—⋀⋁⋀—•

Nick Toby sat at his desk at Lansing Chemical feeling sorry for himself. As a Marine fighting in the jungles of the Pacific he had sacrificed so much for his country. He wanted to keep a "Semper Fi" attitude, but things were not going well. He was beginning to regret the sacrifices he had made. Chemical sales had dropped dramatically after the war, and his commission checks were meager. He and his wife had a three-year-old daughter to raise, and another child was on the way. Nick had made hundreds of cold calls to potential customers, had hustled day and night putting in sixty or more hours a week. And for what? He made more money than this as a college student busing tables.

His desk phone began to ring, and he questioned whether he should even answer it. The odds were good it was his wife complaining they did not have any milk for their daughter.

—◦◦◦—

The phone on the end of the line rang six times before someone picked up. Mary waited, and then a voice.

"Nick Toby, Lansing Chemical."

"Yes, Nick. Mary Morgan at North American Aviation. How are you?"

"I'm, uh, I'm fine. How are you?"

"Great. Hey, I was looking over this product brochure you left me on diethylenetriamine. Do you guys still make this stuff?"

"DETA? Well, we were about to close down production for lack of . . ."

"I want to order some. What did you call it?"

"We refer to it as DETA."[5]

"I just need a small amount to start. But if it works out, we may need quite a bit more."

"Okay. How much do you want?"

"I'd like to start with an order of four."

"Four. Four pounds?"

"No, no. Four tons."

"Four tons."

"How long will it take to deliver it to our test facility in Southern California?"

"I, uh, I estimate about two weeks."

"Any chance you could get it here sooner?"

"I'll prioritize it. We will ship it as fast as possible."

"Thank you. I'm putting a signed purchase order in the mail now."

Mary hung up the phone and ran off to find Tom Meyers.

⌁⟋\\\\⟍⌁

Nick Toby set the receiver back on his phone, stood up, raised one-and-a-half arms in the air, and shouted.

"YES!"

⌁⟋\\\\⟍⌁

"Kinda ballsy to order four tons of propellant before we even know if it'll do the job."

Bill and Toru were sitting in their desk chairs, facing Mary, who was standing at the chalkboard, writing formulas.

"I already figured it out in my head," she responded. "There's no doubt this will work. All we have to do now is calculate the ideal mixture. Ninety percent UDMH, 10 percent DETA? Seventy percent, thirty? Just a matter of number crunching to find the best ratio. Take out too much UDMH and we lesson the performance. Fail to mix in enough DETA and we won't have a high-enough density. My guess is the ideal ratio is going to be around sixty/forty."

"You don't mind if we calc it out," said Toru. "You know—just to make sure."

"It's all yours. Get to it."

Bill and Toru returned to their desks.

Two days later, Mary arrived at her desk just after 8:00 a.m. The previous evening, Bill and Toru had stayed late to work. They had left her a message written in large script on the chalkboard: "60/40 is correct. We love the monkey cage."

Mary was in the cafeteria, trying to choose between yogurt and cottage cheese when Bill Webber ran in. The first thing she noticed was that he appeared to be sweating.

"I found you." He took a moment to breathe. "They approved it."

"Approved what?"

"Your propellant formula. The cocktail. What else would I run over here for?" He took a couple more breaths. "Tom Meyers just stopped by our work station. The mucky-mucks upstairs have approved it for testing."

"Really. They're actually going to mix up a batch and load it in a tank."

"And test-fire it. Tom has already ordered an engine be sent to the Hill. Your little cocktail is going to be run through a turbopump, shot into a combustion chamber, mixed with LOX, and ignited into one helluva fireball. Great news, huh?"

Mary nodded, more preoccupied with her lunch choices than Bill's news.

At that moment, Toru walked through the door and joined them. "Did you tell her?"

"Yep."

"There's just one thing," Toru added. "They want us to name it. They don't want to call it 'it' anymore."

Mary thought about that for a moment, then said, "Tell them we are going to call it 'bagel.'"

"Bagel?" Bill and Toru began to think of what the chemical connection was to "bagel."

Moving down the cafeteria service line, Mary grabbed a container of cottage cheese.

"I don't quite understand," said Bill. "Why are we going to call it 'bagel'?"

"That way," Mary explained, "we can say the Redstone rocket is powered by LOX and bagel."

Bill and Toru were unsure what to say. Was she serious? Was she kidding?

Toru cleared his throat, "If that's what you want us to tell them, then that's what we'll tell them."

Mary grabbed a container of yogurt, looked at it, then put it back. "I usually don't eat in the cafeteria. What's good here?"

"I don't know," said Bill. "Try the bagels."

Once Tom Meyers and the rest of the management team were confident that the new propellant cocktail would work, Bill and Toru were immediately shifted to other assignments in other parts of the Research Group. From this point on, Mary would carry the ball on her own.

By the time the four-ton shipment of diethylenetriamine arrived at Santa Susana, management had decided to pass on the "bagel" suggestion and named the new cocktail "hydyne." Its theoretical density was 0.8615 grams per cubic centimeter, about a 10 percent improvement over alcohol. And its expected specific impulse was 310—also a 10 percent improvement.

On paper, everything was looking up.

19.
SMOKE AND FIRE

"In the 1950s nobody in America knew how to build a rocket. If the Germans had told us to turn around three times and bow to the test stand, we would have done it."
—BILL VIETINGHOFF, SANTA SUSANNA TEST ENGINEER, RET.[1]

Twenty-five million years ago during the Oligocene epoch, a section of land emerged from the sea, pushed upward by immense tectonic forces. The newly exposed earth hardened into a mixture of shale, silt-stone, and conglomerate, with a chico sandstone overlay. For many millennia, the rocky exposition was methodically sculpted—systematically carved by wind, rain, and the occasional volcanic cataclysm. Nature's hand carved the rock strata obliquely, creating mushroom-shaped sandstone formations, a labyrinth of narrow valleys, and a crosshatch pattern of rounded ridges. It was the sculpted sandstone mounds that would give these mountains their peculiar geologic style—a series of hills buttressed by abrupt masses of andesitic and basaltic material and covered with castellated blocks of numerous shapes and sizes. Amongst these rounded crests grew scrub, brush, and sage grass, all competing for limited space and soil. For millions of years it stayed that way, disturbed only by the elements and a wide range of animal life.

Humans arrived eventually, of course, beginning with the native Chumash population that survived off the land for several thousand years. Still, the valley with the low-lying sandstone knobs changed very little.

While all these surface events were occurring over the millennia, something very significant was happening below. Millions of years of accumulated zooplankton and algae had compressed under intense heat and pressure, eventually morphing into an entirely new substance: crude oil.[2] It was the most coveted type of crude—light, untainted, and easy to reach, and its discovery caused the modern world to intrude. It was in 1865 that a Mexican deer hunter named Ramon Peria stumbled across a surface seep of a thin, green oil and staked his claim. Ten years would pass before commercial drilling could begin, but when it did, the investors would claim ownership of the first successful oil well in California,[3] providing fuel for untold thousands of vehicles, from Model A's to Cobra Mustangs.

They called the place Simi Valley.

Despite the importance of the discovery, it was not crude oil that would make Simi Valley famous. Rather, its modern notoriety could be traced to events that began in 1947. That year proved to be a key moment in the history of the sandstone hills. No sooner had World War II ended than the war for technology began. Forty miles southeast of Simi Valley, in the Los Angeles suburb of Downey, a man by the name of Dutch Kindelberger had built a company and called it North American Aviation. A major supplier of fighters and bombers during the war, Kindelberger and his company needed to find peacetime products in order to survive.

The invention of the atomic bomb had created a discussion over unmanned delivery of this new megaweapon. The Germans had been pondering a solution back when the atomic bomb was mere theory. After Hiroshima and Nagasaki, both the United States and the Soviet Union were working the same problem—in the Atomic Age, how does one deliver such a massive bomb to the enemy without risking pilots and planes? And so the idea for an intercontinental ballistic missile was born.

North American Aviation was soon garnering government contracts for large booster rockets, and business got very busy very quickly. Kindelberger had found his peacetime product, and he had found it in the fiery heat of the Cold War.

Soon, thousands of engineers and technicians were ensconced in buildings the size of blimp hangars, toiling away behind 400-pound steel desks. These men—and they were all men—were designing the new motor of the modern age: the liquid-fuel rocket engine. From satellites to cell phones, it was an invention that would change everything for everyone. The new generation of engineers had learned from the Germans, having borrowed technology from their defeated European colleagues; men—and they were all men—so conveniently and secretly imported into the States by the US government. From meticulously kept German records and testimony, the engineers knew their rocket engines would have a high degree of unreliability. "Barely controlled bombs," was an early description.[4] But the cavernous hangars of Downey were too close to civilization to test the products they were making. A safer and more secluded place would be needed, and so a small army of supervisors fanned out like military scouts to find it. The test site had to be remote—far from homes and all the many pleasant accoutrements of civilization that must be protected. This requirement, however, was as much for secrecy as safety. To work on post-war high technology projects, engineers and technicians would all be required to pass a rigorous background check and be eligible for top secret security clearances. Even so, safety was a necessary concern. Large rocket engines were so loud they could, in less than a second, vaporize a man's eardrums or disintegrate a pane of living-room glass. At an average of 180 decibels, large rocket engines were the loudest devices yet created by human hands. Since permanent hearing loss could occur at a mere 140 decibels, even with good ear protection, the engineers knew they needed seclusion—preferably seclusion that was encircled by earthen or rocky berms to muffle and subdue the powerful roars that would be an almost-daily occurrence.

In a raised, boulder-strewn bench 300 feet above the Simi Valley floor, in a fortress-like cirque of those sandstone knobs, engineers and architects found that seclusion. Scattered at random throughout the cirque, a dozen rocket engine test stands were built. Each one was a behemoth the size of a Beverly Hills mansion. They were constructed of thousands of steel beams, wires, and platforms—boxy architectural wonders of anti-aesthetic practicality. Concrete blockhouse control rooms were scattered throughout the area—about one blockhouse for every three test stands, constructed half a mile away for safety. The facility took less than a year to become operational, upon which some overpaid supervisor used a small mountain range to the south as inspiration for its name: the Santa Susana Field Laboratory (SSFL).[5]

Most of the engineers who worked there, however, simply called it "the Hill."

The men who designed and built the SSFL were confident they had achieved the goals of safety, security, and seclusion. No doubt many a company supervisor patted himself on the back, saying, "Forty miles from civilization will make this place completely safe." Yet no sooner had the first bolthole been drilled into the first steel I-beam than real-estate developers began eyeing the orange groves just to the east, a vast swath of flatland named after Mission San Fernando Ray de España: the San Fernando Valley. In short order, the stampede began, commencing in the area now known as Pacoima and steadily working its way west. The onslaught of thousands of three-bedroom, two-bath people-boxes pushed, like a tsunami in slow motion, inexorably westward.

The mad dash of development took only a few years to accomplish; its swath of destruction complete and absolute. Before long, the armies of homebuilders had reached the valley's western terminus, leaving behind not so much as a single citrus tree. It was at that terminus where the backbone streets of Victory, Sherman Way, and Roscoe Boulevard ended—stopped cold at the point where the cheap, buildable land ended and the bouldery terrain began. It was here the ancient Oligocene sandstone pushed itself through the rich soil and began its ascent, rocky and barely passable, into the southland's bronze ozone sky.

●─\/\/\/─●

On the morning of January 5, 1955, the quiet solitude of the sandstone knolls was interrupted by the purr of an in-line six, sky-blue '53 Chevy heading north on Topanga Canyon Boulevard. Ascending gently in elevation, it left the shoulder-to-shoulder world of flatland suburbia in its rearview mirror, then turned left onto a winding, two-lane blacktop called Santa Susana Pass Road.

Mary was sitting up front. She was thirty-four years old now and showed little resemblance to the rag doll from Ray. She was still a short, petite brunette, of course, but now she could afford to take care of herself: makeup, better shoes, a top quality pair of glasses to correct her near-sightedness. Still, growing up on an impoverished farm had instilled in her a lifelong penchant for frugality—all her clothes were still home-sewn.

Mary had never met the driver, or either of the two men in the back seat. They were all rocket test specialists who spent most of their time on the Hill—the kind of techies whom the engineers rarely associated with. She had great respect for these men, since they were the ones who took most of the big risks in the rocket-engine business. These were the men who handled and loaded the propellants—exotic chemicals so hazardous they could freeze an entire human body in seconds, fill their lungs with deadly toxins, strike them with two dozen different cancers, or melt the flesh right off their bones. These were the adrenaline junkies, the risk takers, the aerospace equivalent of mountain climbers. And they were a far different group. The engineers she worked with were very professional, of course, but they were frat boys— the kind of men who were quick with a joke, easy with a compliment, loud and raucous at parties, and generally fun to be around. Most of them, anyway. But the propellant technicians were unlike anyone she had worked with before. None of them said very much during the drive. They were quiet, thoughtful, sober. Most of them, anyway. They talked very little, and when they did talk, it was in whispers. Since leaving the office, Mary had tried several times to start a conversation, but her attempts had led nowhere. These were men with a big job, who had no time for small talk. This lack of conversation only added to her anxiety. Today was going to be one of the most important days of her life. If they were successful, it would also be one of the most important days in modern American history, though almost no one would know about it. Government and company regulations regarding secrecy were as hard as chrome molybdenum steel, and just as unbending.

Mary looked skyward, searching the sandstone knolls on both sides of the road. She had heard that North America's largest bird sometimes made its home in these hills.

"What are you looking for?"

She was surprised—one of the men in the back seat had spoken an entire sentence.

"Condors," she replied. "California condors."

"Oh yeah. They're up here."

That got her attention, and she turned around to face him. His name was Frank, and he had the face of a sailor: sunburnt skin, chapped lips, a neck with too many wrinkles for a man his age.

"Have you ever seen one?"

He shook his head. "But they're around. The Feds come up here once in a while looking for them. Biologists. People like that."

"I saw one once." The driver had come alive, too. "We were doing a test on the NAVAHO engine a couple years ago. A condor circled over the blockhouse compound for about twenty minutes, then flew away."

"If any of you see one, please point it out to me. Okay?"

The three men nodded, then resumed their passive, nonverbal composures.

A minute went by, and they arrived at the turnoff to the SSFL. They would now be off the public highway and on federally restricted land. The driver turned left and arrived at a guard gate. Each individual in the car showed their badge and pass to the uniformed guard, who then made some notes on a clipboard. Then they drove on.

It would not be long now.

Mary grew more nervous, and her breathing became rapid as the front gate appeared around a curve. She swallowed hard and tried to calm her nerves. She removed a pack of Winstons from her purse and tapped out a cigarette. When the driver saw this, he wagged an index finger in front of her face and pointed to a large roadside sign not far ahead.

The sign read, "NO SMOKING BEYOND THIS POINT."

Mary nodded and put away her cigarettes. For now, nicotine was out.

As the car moved past the sign and through the gate, she pondered some of the horror stories her supervisors had related during training. In the early days of liquid-fuel rocket testing, there were many unknowns: accidents and unforeseen incidents happened with regularity. One day, a technician was pressurizing a small sphere. The regulator he was using malfunctioned and indicated a much lower than actual pressure. The pressure in the sphere built up to a point beyond its capacity, and it exploded, blowing the technician's right arm clean off.[6] Another incident involved a technician who had loaded liquid oxygen into a propellant tank just prior to a test. The loading procedure took much longer than expected, and all the while the oxygen vapors that naturally vented and swirled during the operation saturated into his cotton clothing. Hours later, long after the test had been completed, the man lit a cigarette and was immediately immolated in flames. He died on the way to the hospital.[7]

Mary felt a tap on her shoulder. She turned around to face the man seated next to Frank. His ID badge said his name was Roger. His prior experience with rocket propellants was clearly in evidence; he was missing two fingers on his right hand, and his left forearm was heavily scarred.

"Excuse me," he said, leaning forward. "This propellant we're testing. I have a question."

Mary was happy to finally have an intelligent conversation.

"According to my paperwork, it's a mixture of unsymmetrical dimethylhydrazine and diethylenetriamine."

"That's correct."

"Are you sure they're miscible?"

"Yes, they're completely miscible."

"You're certain."

"Yes, we've already tested for miscibility."

"And there were no reactions at all."

"None."

"No temperature fluctuations, no pressure changes."

"None."

Roger grunted, like he was not totally convinced, then leaned back in his seat.

Mary felt lucky to be here, especially since her participation had originally not been authorized. Once the maiden test-firing of hydyne and been scheduled, Mary just assumed she would be allowed on the Hill to watch it. The company would be testing her invention, and in her mind it only made sense for the inventor to be present. Mary's written request for an authorization and pass, she felt, should have been routine. But the next day she received a visit from Tom Meyers, who gently apologized.

"Necessary personnel only," he informed her. "We engineers don't get to visit the Hill very often. We're the ones who design the stuff, but a different crew does the testing."[8]

"That's outrageous," said Mary, her blood pressure quickly rising. "I invented this propellant! I need to see the test—to watch the results."

Tom shook his head. "All you need to see is the data. Within twenty-four hours of the test they'll bring us a box full of charts and graphs with the results. If you're really an engineer, like Webber keeps insisting, you should understand that what you need is data. The smoke and fire of the test may be exciting to watch, but it's meaningless to an engineer."

Of course, he had been right. There was no real reason for her to be present on the day they conducted the first full-up test of her propellant. All she really needed was the data—the thousands of numbers that would pour forth from the chart recorders indicating exhaust temperature, chamber pressure, flow rates, thrust, specific impulse, and so on. Numbers and data. What else was there?

Still, the billows of angry smoke, the torrents of high-speed flame, and *Tyrannosaurus rex*–like roar of the firing would have been great fun. Really great fun—so she decided not to let the issue rest. She pestered her supervisor endlessly to be given permission. When that didn't work, she pressured his supervisor, and then *his* supervisor, climbing the ladder of command rung by rung until she received the proper answer. Finally the word came down from upper management: "We're tired of being pestered by this lady. Just let her go!!"

●─⋀⋀⋀─●

Mary would never forget the first time she went to Disneyland. There was a moment—a brief moment of almost-spiritual transference. The kind of moment whose image glues itself to some secret spot on one's brain cells. For Mary, that indelible moment was the instant she crossed through the turnstile for the very first time, moving from the dreary, gray, asphalt-parking-lot world and into the vibrant, colorful, dreamlike world of Walt Disney's imagination. From night to day, from hell to heaven; it was a moment in time that would always stand out in her memory. Today turned out to be just such a day.

There was a parking lot a hundred feet from the main gate. Their car pulled in and stopped, and all four got out.

The Santa Susanna Field Laboratory consisted of 2,800 acres of knobby hills, buttresses and bowls.[9] Had this been the Old West, there would have been dozens of perfect hiding places for cattle thieves and bank robbers. By the time the 1950s rolled around, there were a hundred buildings, blockhouses, and test stands scattered over those 2,800 acres, all well hidden from each other in the many deep topographical undulations. To get from one place to another, it was best to catch the "Santa Su Bus," a converted World War II vintage ambulance that ran a circuitous route throughout the compound twenty-four hours a day. If you missed the bus, you had to walk.[10]

Today, they missed the bus.[11]

Mary's three companions started walking up a narrow asphalt road. She had not worn her best walking shoes, but she did her best to keep up. Fortunately her love of hiking and backpacking had kept her in shape, and when the gradient steepened, she managed to overtake and pass the three men. Chaparral lined both sides of the road, and the smell of sage and mint mixed with the early-morning air to form a sensation of endless possibilities.

Mary walked another hundred yards, lengthening her lead over the technicians, and came to a place where the road leveled out at a small gap in the chaparral-covered sandstone knolls. Reaching the summit, Mary found herself standing above a magnificent bowl of rock and boulders, like a micro valley. Along one side of the bowl stood three mammoth liquid-fuel-engine test structures. Each one was a complicated matrix of steel I-beams, steel rods, steel platforms, steel buttresses, steel stairs, and an endless number of steel rivets holding the superstructure together. Mary had seen a number of grainy black-and-white photographs of these monoliths—three or four 8 × 10 prints were pinned up on the employee-lounge wall back at the office. Because she had seen pictures, she fundamentally knew what to expect. But it was quite another thing to see the test stands live and up close. She had pushed through the turnstile into Disneyland, and the image stole her breath.

And then she saw it: a Redstone A-7 engine mounted beneath one of the test stands. It stood there, patient and still, awaiting its master's instructions. The engine's bell nozzle pointed downward, as in flight position, aiming its business end toward a massive concrete blast deflection ramp.

But this was not just any Redstone A-7 engine; it was *her* Redstone A-7 engine.

"There you are," she whispered to herself.

As Mary stood at the road's apex, observing the test stands and admiring her engine, the three technicians arrived. Two of them walked past her and headed downhill toward the blockhouse. The third man—the one who had driven their car—stopped and introduced himself.

"Hi. I'm Art Fischer."

"I'm Mary Morgan."

"Impressive, isn't it?"

"I want to live here."

Art laughed. He always took pleasure in observing the look of awe that washed over an engineer's face when they saw the test stands for the very first time.

"We have your A-7 mounted on VTS-1."

"VTS-1?"

"Vertical Test Stand One. The very first test stand built on the Hill."

"Anything significant about that?"

"Yeah. We felt it would be good luck."

Mary took a few more steps to position herself for a better view. She heard some laughter, which drew her attention to the narrow road that stretched between the test stands and the blockhouse. A pair of beat-up World War II surplus Jeeps appeared to be racing from the blockhouse to the stands. Both Jeeps were filled to capacity with well-dressed engineers.

"What are they doing?" she asked.

"Racing."

The engines of both Jeeps sputtered and strained like they were on the verge of giving out, providing the vehicles a just-barely-there power level and giving Mary the impression of a competition between two struggling tortoises.

"Some race," she said, shaking her head.

"Just watch."

A few seconds went by, then suddenly the engines of both Jeeps roared to life with a newfound energy and muscle. The vehicles inexplicably accelerated like a couple of high-power drag racers, turning the struggling tortoises into carnivorous cheetahs. Moments later, the Jeeps reached the end of the road and braked to a tire-screeching stop. Far below, she could hear the drivers and riders laughing and whooping like school kids. Mary turned and gave Art a quizzical expression.

"Engineers. Deep down inside they're all frat boys." He pointed to the test stand with the A-7 engine. "We have the liquid oxygen loaded for your test. The tank naturally vents off vapors, creating an invisible cloud of gaseous oxygen in the vicinity of the test stand. So when . . ."

"So when the Jeeps reach that point, their carburetors get a jolt of pure oxygen," she finished, smiling.[12]

"Mrs. Morgan, welcome to the Santa Susana Field Laboratory."

The garage-size blockhouse sat on a wide concrete pad, surrounded on three sides by the Oligocene knolls. Being the first such building at the SSFL, they called it Blockhouse Number 1. Uphill from it, a seasonal arroyo had once trickled spring runoff through the narrow canyon, meandering gently downward till it arrived at the pancake-flat San Fernando Valley. There it would sate the orange-grove fields secretly marked for destruction. The US government, along with its commercial partner North American Aviation, had paved over the arroyo with concrete and asphalt, diverting the runoff into storm drains. Where the water went after it entered the storm drains, no one could remember. The cement, the concrete, the asphalt, the water diversion—it was all transparent testament to humankind's peculiar need to rework nature's art.

The initial early-morning chill had vanished. It was September, and a Santa Ana condition was brewing, blowing the air hot and dry as it forced its way west. Unstoppable and shameless, the Santa Anas blustered up from the valley floor, turning the hillside sage into crisp kindling. Next to the blockhouse, a few hardy strands of fescue and weeds had managed to shimmy their way up through minute cracks in the concrete pad, where they awaited their herbicide fate. As Mary and Art arrived at the blockhouse, a tumbleweed rolled past. Two coffee-drinking, Kent-smoking men in their thirties were sitting outside at a card table, reading books. They wore the standard uniform of an aerospace engineer: white shirt, dark slacks, black leather shoes, dark tie. A black cat was lazing atop the card table, its eyes peacefully closed.

"Hi, Art," said one of the men. "Who's this?"

"Mary Morgan. She's an engineer."

"Actually, I'm an analyst."

The second man spoke up. "What's an analyst?"

Mary had her answer ready. "An analyst is someone who does twice the work of an engineer but gets paid half as much."

Both men seemed to understand. "Got it."

"Be nice to her," said Art. "She invented the cocktail you're burning today."

Mary could see the change almost instantly on the men's faces. Their expressions went from "What on earth is this woman doing here?" to "I'm impressed."

"Really. Well we'll do our best to make sure you have a successful test."

Mary reached over to pet the cat. "Who's this?"

"That's Newton. He's our mascot here on the Hill."

Newton responded to Mary's touch with a loud purr. He opened his eyes halfway to see who was paying so much attention to him, then went back to sleep.

The Santa Anas continued their bluster as the blockhouse clock ticked slowly onward. The firing had been scheduled for 10:00 a.m., but it was almost 2:00, and no one had any idea when the test would occur. Now Mary understood why everyone brought books to read; testing rocket engines was one of those hurry-up-and-wait activities.

"We're having some problems," was all she was told.

Mary decided to take a walk.

The spacious concrete tarmac in front of the test stands was so new Mary could almost feel the exothermic reactions occurring just beneath her feet. The same thermodynamic equations used in rocket-propellant science were also used in calculating the rates of concrete curing.

Mary ignored the AUTHORIZED PERSONNEL ONLY sign and approached the test stand. She wanted to touch the engine, to run her hands along its polished metal exterior. She had never been this close to a large liquid-propellant rocket engine before. Until now, her career had been nothing but paper and pencils and erasers and slide rules and chalk boards and equations and reference books and mechanical Fridens. Indoor fluorescent lighting, wooden and steel desks packed end-to-end, bridge games at lunch, and tedious staff meetings. But here was where everything came together: a real, live engine, twice the size of her little Volkswagen Bug, ready for its close-up. Massive and impressive, powerful and commanding, awesome and awe-inspiring. In short order, this engine would be ignited, its fuel and oxidizer force-fed into the combustion chamber, and it would pour forth exhaust gases so hot they could melt a car to mush in seconds.

As Mary got to within a hundred yards of the test stand, she realized her depth perception had been off. Everything was much larger than it appeared from four football fields away. The engine she had intended to reach out and touch turned out to be mounted thirty feet off the ground. To get closer, she would have to climb the test-stand stairway. She looked up to see where that route would take her, and then her eyes met those of two technicians perched two stories above her. One of them shouted down.

"Miss, you're not allowed in this area without a hard hat."

The second man said, "Actually, you're not allowed in this area even *with* a hard hat. Who are you?"

"Mary Morgan," she shouted back. "I invented this new propellant we're testing today."

The first man pointed to the blockhouse. "You need to be over there. This is no place for a woman in a skirt!"

Then both men turned and disappeared into the metallic labyrinth of girders and beams, guy wires and gussets, rivets and struts.

Mary shouted back, to no one in particular. "Fine! Next time I'll wear pants!"

She took one last longing look at the A-7 engine high above her, then retreated toward the blockhouse. When she arrived, it was obvious someone had called to report her little test-stand sojourn. A soon as she walked through the door, an engineering supervisor pointed to a metal folding chair sitting alone in a far corner of the otherwise-empty room.

"You can observe—nothing more. Sit there. Be quiet. Touch nothing."

The man had a no-nonsense severity to both the quality of his voice and the composure of his expression. Whoever he was, one thing was certain: he was either in charge of the test-firing or at least considered himself to be.

Once Mary had seated herself, the man left, entering an adjoining room through an open door. Mary could see the second room was filled with electronic gear. Another one of those pesky AUTHORIZED PERSONNEL ONLY signs was stenciled above the doorway.

I'm being treated like a visitor.

The loudspeaker boomed. "Three minutes."

Three minutes. Everything Mary had worked for since the day

Betty Manning dropped her off at Ray's schoolhouse all those years ago came down to what would happen during the next three minutes, forty seconds. Three minutes until ignition, then forty seconds for the duration of the firing. The maiden test of the A-7 engine with hydyne would be shorter than the contract required. As a precaution, advancements in large rocket engines were always carried out in increments. Forty seconds today, 155 seconds sometime in the future. Forty seconds of smoke, flame, fire, and scientific triumph. But only if the engine didn't have a propellant leak, system malfunction, chamber burn-through, meltdown, explosion, or detonation.

The loudspeaker: "LOX tank vent valve closed. Fuel vent valve closed."

Mary wondered why they did not refer to her new fuel by its name.

"LOX tank pressure up. Fuel tank pressure up."[13]

From her position in the chair, Mary would not be able to see the test. After waiting all day? Unacceptable.

"Turbopump lube pump on. Turbopump vent chamber open."

All the test engineers and technicians were preoccupied in the control room next door, and no one seemed to be paying attention to her.

"Main LOX valve closed. Main fuel valve closed. Ninety seconds to ignition."

There was a small viewing window ten feet in front of her.

"LOX and fuel-tank regulators set to fifteen PSI."

Mary looked back toward the control room.

"Fire EX system pressurized. Fire EX main valve closed. Fire EX pump on."

She made sure no one was watching her and stood up from the chair.

"Peroxide regulator set to four-fifty PSI. Sixty seconds to ignition."

She started walking toward the window.

"Open fuel tank pressurizing valve. Open LOX pressurizing valve. Forty-five seconds."

A few steps, and Mary was there. The window was a small rectangle of glass about the size of a toaster, four inches thick. It was mounted high, designed for observers taller than a petite woman. Mary stood on her toes. The window was dusty, but it had a clear view of the test stand.

"Igniter on. Twenty seconds."

The man who had ordered her to essentially sit down and shut up had not returned. Mary glanced again to her left to make certain no one was paying attention to her, then turned her eyes again to the test stand. Through the window she could see the vented oxygen vapors whispering round the propellant tanks, like some large, ghostly apparition. Mary stood motionless, eyes fixed on the Redstone A-7 engine.

The outdoor siren began its loud wail—the "ten-second warning."

"Ten seconds. Nine . . . eight . . . seven . . . six . . ."

Mary held her breath.

Here we go.

A few more seconds, then she saw it—a large flame pop out of the engine. Then nothing, then another pop of flame. And then it ignited full-bore, blazing with a stinging light as if from the sun and roaring like a monstrous earthquake.

It ran for three seconds. Then it stopped.

Three seconds. A measly three seconds. Mary stood there, astonished.

"What happened!?" she shouted. "What the hell just happened!?"

"Authorized Personnel" be damned; Mary ran into the control room.

"What was that!? We're supposed to have a forty-second run!"

Art Fischer walked over, an apologetic look on his face. "Sorry. Our safety systems initiated an automatic shutdown. It registered some combustion instability in the chamber."

"What caused the instability?"

All the technicians in the room laughed. Three-finger Roger stood up from his chair and faced Mary.

"You tell us. You're the engineer."

DON'T DRINK THE ROCKET FUEL

**"The NAVAHO rocket was like an Apple II computer;
it was state of the art for about four weeks."**
—G. Richard Morgan, rocket engineer, ret.[1]

F rom the loudspeaker atop the blockhouse came the sound of a bland, robotic, male voice.

"LOX vent valve closed. Hydyne vent valve closed."[2]

Mary stood in the control room. She turned to face the man with the microphone and smiled.

"So you're finally calling it by its name."

Several technicians walked over and shook her hand. Each of them offered her a derivative of, "Good luck." Today was the fourth hot-fire test of the new propellant. Ironically, they had been having more problems with hardware than with the new fuel. So far, not a single test had managed to get a burn longer than thirty seconds without being shut down. The good news was that all the data they collected in the first three tests showed the specific impulse was above the contract specs, averaging out at a respectable 309.[3] But since the army contract required that North American demonstrate three successful full-length static tests of one hundred fifty-five seconds each, they were still at it.[4]

The robotic announcer again held forth. "Lox and hydyne tank regulators set to fifteen psi. Ninety seconds."

The surrounding hills were not high enough to capture such sounds and bounce them back, so the announcement, though loud and firm, landed like a thud, more absorbed than echoed. Irving Kanarek, sitting just outside the blockhouse with three technicians, took a last sip of his coffee, stood up, then added his paper cup to a hundred others nesting inside a fifty-gallon steel barrel.

Bill Webber appeared in the doorway to check for stragglers.

"You guys better get inside."

Irving nodded, then he and the technicians entered the blockhouse.

In the control room, Mary had a pair of binoculars pressed against her eye sockets. She could see the white oxygen vapors venting from the LOX tank, whispering around the steel girders until a Santa Ana gust stole them away. In a little over a minute, a couple of monster turbopumps would mix thousands of gallons of liquid oxygen with a like amount of hydyne to create a hurricane of boiling fire. A 4,000-degree flame would pour from the nozzle and roar from the engine to form a violent pressure wave of quake-like sound.

"Fire ex system pressurized. Main valve closed. Pump on. Sixty seconds."

Mary adjusted the focus on the binoculars. There was something moving on the road between the second test stand, VTS-2, and VTS-1 where the A-7 engine was mounted. It looked like a man, and he was staggering. Then suddenly he fell to the pavement and rolled downhill for about ten feet.

"There's somebody out by the test stands!" she shouted.

"We see him. Looks like Toumey's drunk again. We're gonna hold the count at fifty-three seconds."

Mary was incredulous. "Is he a technician?"

"Yes."

"And you let him drink on duty?"

"Not exactly."

Bill grabbed a set of keys hanging by the door. "Let's go get him."

Mary took the keys from his hand and led the way outside. She jumped into the company-owned Jeep parked next to the blockhouse. Bill followed right behind, taking the second seat. As Mary started the engine, Bill looked at Irving, who was standing at the open door.

"Coming?"

Irving shook his head. "You two can handle it."

As Mary made a U-turn with the Jeep and accelerated toward the test stand, the loudspeaker blared. "We have personnel in the firing area; holding at fifty-three seconds."

The Jeep engine was loud, and Mary had to shout over it.

"Who is this guy?"

Bill shouted back. "George Toumey. He likes to drink the rocket fuel."

—⁓—

Irving watched as the Jeep bounced over the fresh black-asphalt road leading to VTS-2. An engine using LOX/ethyl alcohol was scheduled to be fired from that stand later in the day. Holding his hand to shade his eyes, he could see the technician lying and rolling on the ground. George Toumey was one of the equipment specialists in charge of plumbing the fuel lines from the propellant tanks to the rocket engines. As such, he knew how to tap the fuel lines and drip out a pint or two of alcohol, ostensibly for purity tests. Unfortunately, he kept testing it on himself. George Toumey had been warned many times by management not to drink the rocket fuel, but he just couldn't help himself. Once in a while, when he thought no one was looking, George would tap out a small amount of the propellant, drink it, check for witnesses, then repeat the procedure. Several times. More than once he had gotten so drunk he had to be carried away. Never, however, had it happened so close to a test-firing.

Joe Friedman—one of the engineering supervisors—stepped out of the blockhouse to watch. Several others soon joined him.

"Old George would have gotten a four-thousand-degree ass-kick if somebody hadn't noticed."

"Now *that* would have taught him a lesson."

"If he survived."

"Holding at fifty-three," said the calm, emotionless announcer.

Several of George's fellow technicians had noticed he was missing, and it did not take them long to figure out where to find him. As Mary and Bill arrived at the spot just below VTS-2, three technicians joined them to help lift their sotted friend onto the Jeep. As the technicians ran for cover, Mary put the Jeep into gear, turned the car around, and headed back for the blockhouse.

"If he's done this before, I don't understand why this guy hasn't lost his job yet," said Mary as she shifted to third.

Bill turned to look at the sleeping technician. "I suspect this will be his last day."

The tires squealed as Mary pulled the Jeep up to the blockhouse. Irving and Joe were there to help, and the four of them managed to carry the groaning, moaning technician into the blockhouse.

"Resuming count at fifty-three seconds," blared the loudspeaker.

As the countdown continued, Joe Friedman knelt down next to the sleeping George Toumey and slapped his face a few times to wake him up. It took a few moments, but George finally stirred, opening his eyes and looking into the face of his friend.

"Hi, Joe. Where am I?"

"You're fired."

"Oh."

George Toumey went back to sleep.

Mary was angry that her test had been interrupted by an act so unprofessional. Despite the fact that the test was imminent, she stood outside the blockhouse door with a cigarette, trying to calm her nerves. Bad enough that she was under so much pressure to get hydyne into the Redstone and help launch America's first satellite; the last thing she needed was the George Toumeys of the world slowing her down with their reprobate habits.

Standing beneath the "No Smoking" sign, Mary took another drag on her cigarette. Newton—the compound's adopted cat—appeared out of nowhere, brushed against Mary's calf, then darted off toward the brush to find mice. Mary's eyes followed where Newton had gone, and near that spot she noticed two deer were standing very still atop a boulder, watching her watch them.

Newton had shown up one day during the early construction

of the Simi test facility. The construction workers had adopted him, sharing bites of their sandwiches and bits of cream meant for their coffee. Eventually the builders left, leaving the site well constructed, but cold and lonely. The cat had stuck around, though, showing canine-like instincts—a loyal puppy waiting for its master's return. Soon after California approved and issued Santa Susanna's operational permits and licenses, the engineers and techies moved in. The first group to arrive noticed the cat—how it would show up and beg for food each morning and afternoon. Like the construction workers before them, the engineers adopted him, naming him after their favorite scientist.

Newton returned, and once again ran its fur across Mary's calf. She bent over and gently petted his fur.

Mmeewwww.

—⌁/\/\/⌁—

Mary felt a cold Canadian wind—the kind that pulled with them the dreams of young girls. She was petting Missy, their calico cat, who was nuzzling her leg. Mary's eyes widened in shock, and her mouth opened in astonishment at the sight of the skinny feline. Missy was no longer pregnant; somewhere on the farm, a squeaky litter of kittens was battling for survival.

Mary would have to find the litter and hide the gunnysacks—and fast.

"Come here."

Mary set down the milk pail and picked up the cat, looking around to make sure no one was watching. There were only a couple of places Missy would feel confident about having her litter—either a dark, private corner of the barn, or the crawlspace under the house.

"If father finds out, it'll be the gunnysack for sure."

Unlike their house, the barn was a relatively new and sturdy structure. Michael had built it with money earned from his second vocation as an amateur veterinarian. For reasons they never fully explained, the Norwegian and Danish farmers did not like to care for the medical needs of their own animals, always turning to Michael Sherman when a horse turned sick or a cow became bloated. In the off-season, Michael Sherman made more money as an untrained vet than he did as a farmer. The new lumber was testament to this. Mary lovingly petted the cat as she explored the barn's numerous nooks and crannies.

"Where are your babies, little mommy?"

As Mary explored the barn, she approached a far corner, where a small stack of hay was piled, Missy became fidgety, squirming nervously.

"Is this where they are?"

Missy dug a claw into an arm, extricated herself from Mary's grasp, leaped into the air, and hit the ground running. To Mary's surprise, the cat sped into the field. She gave chase.

Missy had a single eye of bright amber, having lost her left eye years ago during an altercation with a young grey wolf that thought the cat might make good supper. Missy still held the scars from that fight: two long gouges traveled from her left eye socket down to her belly. Now, as she slowed to a careful walk, meandering through the tall stalks of bright-green field grass, she took the occasional furtive glance behind. The girl was following, Missy was sure, though she couldn't see. Instinct told her the girl was not a threat. Still, in the last three years, she had delivered four fine litters, and all of them had mysteriously disappeared soon after birth. Somewhere in the land, there was a monster.

Missy tilted her head and sniffed the air. She was close now. She pushed through the tight columns of grass, stopping briefly to scratch her ears on a sharp rock. There was a sound, and she turned to look.

The girl was close. Missy decided the risk was too great and ran off in another direction.

When Mary saw that Missy had abruptly changed her trajectory, she guessed it was a ruse and kept to the beeline her cat had been walking. Another minute, and the old tractor appeared—almost swallowed by the field grass. Only the torn leather seat and rusted exhaust pipe were visible.

And then the tiny cries became audible. Minute quasi-harmonies riding a high-octave wave, the chorus was both discordant and beautiful. Mary stood next to the tractor—a dead, unburied body of rusting steel wormed out by the elements of sun, wind, and rain.

"Where are you?"

Mary followed her ears and soon found them—six shapeless kits, squirming together like maggots on meat. Missy had nestled them atop a tuft of dried grass inside the engine cavity, where they would be somewhat protected from the weather. Their eyes were still shut, but their mouths had no problem with function. Mary reached in and was about to lift one out when suddenly Missy pounced—jumping out of nowhere onto the leather seat. In Missy's eye, Mary could see the deep concern of anxious motherhood and withdrew her empty hand.

"Don't worry, girl. Your secret's safe with me."

Missy meowed loudly, then leaned in to check on her litter.

Mary placed several tufts of dried grass around the newborns to keep them warm and dry, and Missy did not seem to mind. Mary climbed atop the tractor and stood on the leather seat, surveying the farm in all directions. She could see no one and decided the kittens were safe where they were. Besides, Missy would throw a fit if she tried to move them.

Mary jumped to the ground and ran back toward the house.

As required, all Sherman family members were present at the supper table. Michael at one end, Dorothy at the other. Sisters Amy and Elaine sat on either side of Mary, while the boys—Clarence, Michael and Vernon—sat opposite. The table was set with homemade foods: farm-centric victuals that would be unrecognizable to more-processed generations yet to come. The lukewarm liquid with the unfiltered floaties was milk, the coagulated lump of malleable curd was cheese, the lumpy gobbet of slippery white pudding was butter.

Mary's father pointed a fork in her direction.

"The only homework you need to worry about is home work— milking the cows and shoveling their manure." He stabbed the fork into a slice of carrot and delivered it to his mouth with purpose. "And don't forget to clean the creamer."

Across the table, all three boys were stifling laughter. Mary turned to her mother but could see right away that sympathy was not on tonight's dinner menu.

"Listen to your father."

"Tell Bowman your dog ate your homework," said Clarence. "That's what I used to do."

Mary turned to her mother. "My teacher assigned me the alphabet. I havta memorize it by tomorrow."

"We all have chores to do."

"All the other kids in my class know how to read. I don't even know the alphabet. I'm the only one! The kids think I'm stupid!"

Vernon could not resist. "You *are* stupid."

Dorothy gave her son's head a swat, then turned back to her daughter.

"Obey your father like the Bible says, or I will give your dinner to the boys. Lord knows they need it, with all the work they do." She changed the subject. "Father—we're almost out of lignite."

"I'll take a couple o' the boys into town. Clarence—you stay here. Turkeys and hogs need feeding."

"Yes, sir."

Then Mary's mother said something that made her freeze.

"After you're done with the feeding, I need you to fetch the gunnysack."

"Missy have her kittens?" asked Michael.

"No she didn't! No!" But Mary knew her lie was of no use; somehow her mother had discovered Missy had given birth. The cat's diminished girth had not gone unnoticed.

"It has to be done," her mother answered coldly. "Farm can't support any more pets. They have to go."

Mary looked around the table for support. Even her little sister, Elaine, the one person she could sometimes count on for support, avoided her eyes. Mary teared up.

"May I be excused from the table?"

—⋀⋀⋀—

The ten-second warning siren blared, and Mary returned to the real world.

What am I doing out here?

She ran into the blockhouse and grabbed her binoculars.

At zero seconds on the count, the rocket engine came to life. A tempest of fire, a tornado of flame, roared from the rocket's nozzle at a speed greater than sound. A moment passed, then the shock wave collided with the blockhouse, pounding it like a 9.0 earthquake. Except for Mary, everyone watched the test without much emotion—they had experienced this a hundred times before. It was all very routine— science going about its daily tasks. Other than the ethyl alcohol–loving George Toumey, it was just another day at the office.

Thirty-seven seconds later, the rocket engine cut out.

Data reduction would show the engine was continuing to experience combustion instability. And when the engine was taken apart for examination, they would find part of the injector plate had melted. Still, at thirty-seven seconds, it was their longest and most successful test yet.

Within a minute, the all-clear siren sounded throughout the compound, and the test crew left the blockhouse for the test stand to inspect the results of their work. Except for George's sleeping form on the blockhouse floor, Mary was the only one left. She glanced at his rotund, snoring hulk, then shook her head.

"Drunks," she whispered to herself.

The engineers had left the door open, and Newton ran in to stand next to Mary. But when he realized she had no food for him, he scooted off to destinations unknown, ostensibly to beg from someone else. She stepped to the doorway and watched Newton run into the sage-covered hills.

That's when Mary noticed that the two deer were still standing in the same spot—they had fearlessly stuck around to watch the engine test.

But there was no time to worry about hungry cats, curious deer, or thirsty technicians—there were far too many specific impulse figures to calculate, mixture ratios to work on, test results to analyze. The engineers and technicians could afford the hierarchal employment luxury of hanging around the test stand all day, beating their chests over the latest combustion-chamber instability problem—Mary had work to do. Somewhere in all the data rolling out of the chart recorders were the answers. She would study them, and she would solve the problem.

Now, if only George Toumey would stop snoring.

21.
PUSK!

"I never think of the future. It comes soon enough."
—ALBERT EINSTEIN

Sergei Korolev had sacrificed everything for this day. His youth had been spent (wasted, many would say) daydreaming of space-travel glory. His education and employment had been focused on one course of action. His marriage to Ksenia was over before it started, the pursuit of wish fulfillment overriding any and all spousal affection. His intellect had made him a target of dictatorial purges. His six-year prison sentence in less than humane conditions had brought about a long string of debilitating health problems. Family, friends, freedom, relationships, social activities, vacations, both kidneys—all sacrificed on the altar of single-minded obsession. And every one of his sacrifices could have been so easily avoided if only Sergei had been a simple man with meager ambitions and unremarkable goals.

Of course, such people rarely make history.

The autumn of 1957 arrived, and along with it the International Geophysical Year. Extending from July 1957 through December 1958, the IGY was a one-and-a-half-year period of time in which sixty-seven countries would participate in observing and recording various geophysical phenomena around the world. One of the goals of the IGY was for both the Soviet Union and the United States to attempt the world's first launch of observational satellites. It was not a race or a competition, and in fact the expressed intention of the IGY was that all of its activities be purely scientific, washed pure and clean of all nationalistic and political dirt.

It didn't work out that way.

The massive R-7 rocket had enjoyed three successful flights in a row—an aeronautical first in its own right. The Soviet Union was now the only country on the planet with proven heavy-lifting rocket capability, and almost no one was aware of it. Sergei scheduled the world's first satellite launch for October 6. Then a cable was received from Moscow: the IGY had scheduled a surprise meeting in Washington, DC, for the same date. According to the message, the highlight of the meeting would be a major announcement by the United States about an "American satellite." Normal IGY protocol would require Korolev to be at that meeting, which would force him to postpone the launch of *Sputnik*. The coincidental timing, along with the cryptic nature of the Americans' "announcement," drove Korolev into a state of anxiety and panic.

Why the sudden meeting? Why that particular date? Had the Americans gotten wind of his project? What did it mean? Were the Americans planning a satellite launch? Or perhaps they had already achieved orbit and were intending to announce it at the meeting.

No—there was no way they could have kept such an achievement secret. The announcement, Korolev concluded, must be about an impending launch, perhaps even on that very day. To convene the IGY on the same day the Americans reached orbit would have dramatic flair, the kind of flair the Americans had a reputation for. So concerned was he about the Americans and von Braun beating him into space that Sergei requested the assistance of the KGB. He asked them to find out if an American satellite launch was imminent. The KGB promised to have their spies and agents in the United States investigate.[1]

"We'll get back to you," they said.

Sergei's space-travel fixation had always been more than the simple launching of a satellite. Even though most people were unaware of it, rocket technology had advanced to the point that any good team of engineers, with enough money and resources, could place a satellite into orbit. That's what was so scary—satellite-launch technology was low-hanging fruit, yet no one was reaching up to grab it. So putting a satellite into orbit was not the goal. In Korolev's view, the real goal was to launch the world's *first* satellite. Nothing else would matter. The first person to accomplish that milestone would not only make history but would achieve immortality. Whoever was number two would be quickly forgotten, ignored by history as nothing more than a statistical asterisk. Sergei Korolev refused to be a forgotten statistic.

The KGB sent him a cable: there was no immediate effort on the part of the Americans to launch a satellite. In fact, there seemed to be no feeling of urgency anywhere in government or among the citizenry for such a thing. To Korolev, this report meant that his fears were justified; whatever the Americans were up to, they were keeping it top secret. For far-sighted scientists like him, it was simply too far-fetched to think the Americans and their leaders would be more interested in the latest baseball scores than seeking the glory of space. The KGB assessment, he concluded, was wrong. Time was a monster—the kind of monster that gobbled up dreams rather than flesh. Time was Sergei Korolev's archenemy—it was always against him, always fighting him, always there to conquer and destroy his dreams. Time was the devil.

Korolev was a brave man. He had stood up to the Soviet military leaders and won. He had faced down the politburo and come out on top. He had overcome the doubts of Nikita Khrushchev and gained his faith and trust. Such bravery was as rare as moon rocks in 1957 Soviet Russia, and yet Sergei Korolev was not a man without fear. There was one thing that kept him awake at nights. One thing that invaded his dreams and turned them into nightmares. One thing more fearful than the torture chambers he had endured in the concentration camps. The only thing Sergei Korolev was deathly afraid of was being number two. If there was to be a major announcement at the October 6 meeting, Sergei Korolev wanted to be the one who made it. It would be he who would wear the laurels of glory. It would be his picture on the front of every major newspaper and magazine. It would be his pen that signed the autographs. It would be he universally praised for his technological brilliance. Despite a dearth of evidence that the Americans had any sort of launch attempt in the pipeline, Korolev chose not to take any chances. In the race for space that only he and Wernher von Braun seemed to know even existed, two days was not two days; it was an eternity. And so one cold Tyuratam morning, Sergei Korolev called his colleagues together and declared that he was moving the satellite launch date up two days to October 4.

As representatives around the world began buying their plane tickets to attend the IGY set for October 6, Korolev and his team sent along apologies and phony excuses for their absence, then prepared for launch.

Only later would Korolev discover how accurate the KGB reports had been.

The R-7 rocket was assembled first in an upright position at a specially built hangar, then slowly towed 1.5 kilometers to the launch area using a tractor platform. Their indoor vertical preassembly technique would eventually be adopted by the United States and NASA, but as with so many other large-rocket details, Korolev and the Soviets had found the wisdom in it first.

The chief designer walked in front of the rocket-carrying flatbed, as if leading it to its destination.[2] It took the large transporter almost an hour to travel the short distance to its launch area.[3]

Once it arrived, and all the systems had checked out, authority for the launch passed from the civilians, headed by Korolev, to the military, commanded by Colonel Aleksandr Nosov.[4]

"One minute to go," said Nosov. "Key to launch."

Lieutenant Boris Chekunov inserted a key into the command console. This key controlled a circuit breaker for the firing circuitry.

"Key on," said Chekunov.[5]

Nosov ordered the engine feed lines to be purged with gaseous nitrogen in order to flush out any residual fuel or oxidizer that might be remaining from the propellant-loading process.

"Key to drainage," said Nosov, and Chekunov responded by shutting off the liquid-oxygen relief valves. Two minutes passed, and Nosov gave his final order.

"*Pusk!*" A single word, meaning "launch."

Chekunov now pressed a button that triggered a series of automatic events, like a line of falling dominos. Valves opened, allowing gaseous nitrogen to pressurize both propellant tanks; the umbilical connections, mostly electrical lines, were retracted; the rocket was placed on internal battery power; the turbopumps were actuated, forcing the propellants into the engine's combustion chamber; igniters inside the combustion chambers were switched on, torching the now-mixed propellants and causing a brilliant upside down volcano of fire.

"Primary stage!" shouted Nosov as the turbopumps methodically throttled up to full power.

The rocket began a ponderous upward climb.

"Liftoff!"

22.

THE DUTCHMAN COMETH

**Success consists of going from failure to failure
without loss of enthusiasm.**

—Winston Churchill[1]

The failure of American and international leaders at all levels, both civilian and military, to understand beforehand the significance that a maiden satellite would have on Earth's human population turned out to be one of the most spectacular political blunders of all time. For years, a small handful of engineers, led by Wernher von Braun, had pressed President Eisenhower and his government about the urgency of moving satellite technology off the drawing boards and into space. But the president had always been dismissive of their warnings. The political climate would have been just as bad in the Soviet Union had not Stalin and his military leaders been so hungry to build an arsenal of ICBMs. Only a small number of countries had the engineering know-how to launch a satellite, and each and every one of them was crippled by institutional myopia.

Soon after *Sputnik*, a Gallup poll determined that 50 percent of the US population considered the Russian achievement a "serious blow to US prestige."[2] In an interview he gave in 1998, President Eisenhower's staff secretary, Colonel Andrew J. Goodpastor, said the Americans seemed appalled by the president's indifference, saying, "It really created great anxiety, almost panic within the United States."[3] In their book *Project Vanguard: The NASA History*, Constance Green and Milton Lomask describe the weeks of post-*Sputnik* as "a period of mental turmoil and vocal soul-searching."[4]

The American media was much more brutal. *U.S. News & World Report* and *Aviation Week* began a series of articles highly critical of America's new laggard image in science and technology.[5]

On the evening of *Sputnik*'s launch, there was a party at the officer's club in Huntsville. When Wernher von Braun arrived, he took General Medaris and US Defense Secretary Neil McElroy aside for an urgent discussion.

"If you go back to Washington tomorrow, Mr. Secretary, and find all hell has broken loose, remember this: We can get a satellite up in sixty days."[6]

There was a stunned silence. Most everyone was still under the assumption that Vanguard was destined to become America's first foray into orbit. Von Braun's declaration seemed desperate and presumptuous, given, as it were, in front of one of the President's cabinet members. But von Braun wasn't done speaking.

"Vanguard will never make it. We have the hardware on the shelf. For God's sake, turn us loose and let us do something!"[7]

Secretary McElroy and General Medaris did indeed go to Washington to plead von Braun's case, but Eisenhower turned them down flat.[8] What was lost on all of the space boosters' satellite dreams, however, was what was going on in the minds of common Americans. If the communists, America's Cold War enemy, could launch a satellite overhead, that meant they had the capability to hit New York or Los Angeles with an H-bomb.[9] Americans did not like being second in the newly declared Space Race, but they hated living in fear even more. Overnight, two new phrases entered the English lexicon: *fallout shelter* and *bomb shelter*. Fearing a nuclear holocaust, people started building them in their backyards.

Most everyone in the United States agreed that America needed to leap into the fray and show the world its technological superiority. Now that the informal Space Race had been unofficially declared, it was assumed by Americans everywhere that their country would demonstrate its true mettle and capability. Soon after the launch of *Sputnik* on October 4, 1957, the VOG (Vanguard Operating Group) announced that on December 4 it would launch the Vanguard TV-3 rocket from

Cape Canaveral. This launch was intended strictly as a test; there was never any claim or pronouncement that TV-3 would be America's official first satellite launch. Misinformation and exaggeration, however, increased public awareness to the point that the VOG decided to "go for it" and placed a token electronic package atop the rocket "just in case" it happened to go into orbit.[10] For this reason, hordes of people, reporters, and cameramen descended on Cape Canaveral. Expectations were high, and completely misplaced.

The launch was beset by technical and weather delays that pushed liftoff from December 4 to the 6th.[11] On Friday, at 11:44 a.m., the launch sequence for Vanguard TV-3 began. Pyrotechnic igniters in the engine were started, the propellants—liquid oxygen and kerosene—were forced into the combustion chamber, and the engine started with a typical large-liquid-rocket-engine roar. The rocket lifted off the pad, and for about two seconds all looked well. Then someone shouted, "Look out! Oh, God, no!" The rocket suddenly lost thrust and descended downward, back toward the launchpad. As the rocket crumpled on its own weight, the propellant tanks ruptured, the fuel and oxidizer mixed and ignited, and Vanguard TV-3 exploded in a massive fireball. One of the engineers, Kurt Stehling, would later describe the vision of what he saw from the blockhouse window. According to Stehling, it seemed "as if the gates of hell had opened up."[12]

With so many journalists on hand for the launch, it was inevitable that the explosion would play out in the newspapers and television for days and weeks on end. As if the Soviet's launch of *Sputnik* were not embarrassment enough, now America had on its résumé a spectacular rocket cataclysm—all on the six o'clock news for everyone to see.

The dichotomy of that one incredible Soviet success matched against America's monstrous failure changed everything.

—•◦◦◦•—

James Howard "Dutch" Kindelberger was born in Wheeling, West Virginia, on May 8, 1895. His father was a steelworker, a position

James worked at for a short while before enrolling at the Carnegie Institute of Technology in 1916. The United States entered World War I the following year, and James joined the army, serving in the Aviation Section of the Signal Corps. He became a pilot instructor and was soon enamored of anything that had to do with aircraft. In 1919, he married Thelma Knarr, and a year later, became chief draftsman and assistant chief engineer with the Glenn L. Martin Aircraft Company in Cleveland, Ohio. Five years later, he accepted a position as chief engineer with Douglas Aircraft in California, where he pioneered development of large passenger planes, the DC-1 and the DC-2. In 1934, Kindelberger became president and general manager of General Aviation, later renamed North American Aviation, Inc.[13]

Though Dutch, as he would come to be known by both friend and foe, was well regarded for his engineering talents, he was legendary for his marketing acuity. During World War II, North American Aviation was turning out more than five hundred planes of various designs per month for several branches of the military. At that time, the United States Army Air Corps issued most aircraft orders. Not until 1947 would the Air Force become an official separate branch of the US military.[14] One day, Dutch was summoned to the USAAF headquarters in Dayton, Ohio, for the purpose of discussing a new airplane the generals wanted to buy. Kindelberger knew they were heavy with bombers but light on fighters, so the week before he was to meet with the USAAF brass, he quickly drew up plans for a new advanced fighter plane. Then he packed the blueprints into a suitcase and piloted his own plane to Dayton.[15]

The morning of the big meeting, it was raining heavily. Fortunately, Kindelberger had brought his raincoat, inside of which the valuable advanced fighter plane blueprints were safely tucked. As he entered the office building, he shook the rain off his coat, and he was soon ushered into a room full of generals and colonels, all sitting around a long, rectangular conference table. For visual effect, Kindelberger laid the blueprints out on the table as he stood before them.

"So," he began. "What can we make for you gentlemen?"

The response made Kindelberger stop breathing.

"We want a new bomber. All the bombers we have are too small. We need a bomber that can fly farther and transport a much greater payload."

As soon as he heard the word *bomber*, Kindelberger had casually rested his raincoat atop the blueprints, hiding enough of the design so that no one would be able to tell what kind of plane it was.

"Yes," said another general. "We need a much bigger bomber. We hear your competitors are already working on designing such a bomber, so we're not sure if we should even be talking to you."

"This is your lucky day," said Kindelberger. "We've spent the last six months secretly developing a large bomber just as you have described." Pointing at the obscured blueprints he said, "And unlike my competitors, our design is ready to go."

Dutch Kindelberger left the room with the largest order of bomber aircraft he had ever received. On the flight home, he began designing the new plane in his head.

Tom Meyers's office was raised ten feet off the engineering floor and positioned so that from his windows he could see every last slide-rule puller in the department. When his phone rang with the warning, he had not been at those windows, sitting instead at his desk pouring over the latest engine-performance data.[16]

"He's in the building. He's not happy. Make sure your desk is clean."

He recognized the voice of Audra, one of the lobby receptionists. She had barked off those three quick sentences, then hung up the phone.

He's in the building? Who *is in the building?*

Obviously they were receiving a visit from some VIP—not an unusual occurrence. An army general, or perhaps even a congressman. As Tom stood up, Irving Kanarek ran into his office.

"He's in the building!"

"Who?"

"Mr. Big. Mr. Mucky-Muck himself."

Tom was confused. "Sam Hoffman?"

"No—no! The *real* Mr. Big! The legend: Dutch Kindelberger." Then Kanarek left.

Tom stepped quickly over to the windows. Kindelberger was easy to spot. As he marched toward Tom's office, the founder and commander of North American Aviation left a visual wake of turning heads, moving bodies, and fear. And they had good reason to have that fear. Kindelberger was famous on Wall Street as the inventor of the "Five Percent Rule"—a policy that required 5 percent of all company employees to be fired every year, whether they deserved it or not. Kindelberger's theory was that such a rule made the other 95 percent work harder. Sadly, it often succeeded.[17]

Tom struggled to consider why Kindelberger would be paying him a visit.

In ten years, Dutch has never once visited this plant. Now he's here, and heading my way.

Tom rushed to clean his desk, a job he managed to get half completed before his door opened and the Dutchman entered. Kindelberger was wearing a black suit with a white shirt and a blue polka-dotted bow tie. His hair was receding, but most of it was still there, stark in its whiteness.

"Hello, Tom."

"Mr. Kindelberger." Still holding an armload of files that he had intended to hide somewhere, Tom stood up.

"Mind if I come in?" said his visitor.

"It's your company, sir. You can go wherever you want." Tom set down the files.

"I know. But courtesy, Tom. Never forget it. Courtesy goes a long way."

Kindelberger took a commanding stance in the middle of the room.

"Let me ask you something, Tom. Do you have any idea what it's like getting a phone call from the president of the United States?"

"Can't say I've had the pleasure."

"It's not something you'll ever forget. Trust me."

The white-haired gentleman began walking the room, examining a display of photographs on the office wall. Some of the photos were of successful rocket test-firings, some depicted famous NAA aircraft from the war. Then he arrived at a photo of Tom and his family.

"How's the wife and kids?"

"Just great. We're all doing fine."

Kindelberger turned and stepped directly in front of Tom's desk.

"That's wonderful. Having a family is a wonderful thing, isn't it?"

"Yes, sir; it is."

"And being able to support them; that's a wonderful thing, too."

"Absolutely. So, I take it the president called."

"I got two phone calls yesterday. One from Dr. Wernher von Braun, and one from President Dwight D. Eisenhower."

Tom shook his head in amazement. "Wow. Wernher von Braun."

"Ya know, in World War II, our company built thousands of planes and bombers designed to kill the sonovabitch. Now we work for him. Half the contracts this division receives have his signature at the bottom."

"Strange how life works out."

Tom's golf bag was leaning against the wall. Kindelberger lifted a club halfway out, briefly examined it, then put it back down.

"There's a lot at stake here, Tom. I've given my heart and soul, my life, and every penny I have to build this company. We have a great reputation with our customers. Now that reputation is on the line."

"I know that, sir. And you know we've always put out a quality product."

"Yes, of course. Still, there's this . . . girl." He drawled out the word *girl* in a manner barely masking his disdain. During the war, Kindelberger had depended on thousands of women to build his fighters and bombers, but it wasn't something he approved of. In Dutch's world, a woman's toolkit included fry pans and spatulas, not rivet guns or arc welders.

"What's her name—the one you got on the propellant contract."

"Mary Morgan."

"Mary Morgan. Right. How's she comin' along?"

"Real good. Making steady progress."

Kindelberger had removed a nine iron and was examining it closely, like an art critic would eyeball a Paul Cézanne. "Don't bullshit me, Tom."

"Okay, okay. Look, she's struggling; but remember—not even von Braun's best engineers could solve this problem."

Kindelberger let the nine iron slide back into the bag, then he began touring the office again.

"Ya know what that goddamn satellite is doing, Tom?"

"Well I know its orbiting."

"It's beeping. *Beep-beep-beep.* And it's doing it on a short-wave frequency anybody in the world can tune in to and listen. Here we are trying to keep the world safe from communism, and every kid in America is listening to their *beep-beep.*"

A shelf unit had a number of memorabilia and tchotchkes, and he began picking them up and examining each one.

"We're in an undeclared space race here, Tom. Russia puts a satellite into orbit one day, and the next day America is some snot-nosed pooper grabbing his mommy's dress and crying to have his diaper changed. I always said: first country to put a satellite into orbit will be on top of the world. Ya can't buy that kind of prestige with money, Tom. You buy it with grit. You buy it with risk. You buy it with balls. The whole world, I said, the whole world is going to look to that country as a leader. And the company that puts them there—they're gonna get a helluva lot of contracts. We don't want the world doing business with the Russians, do we, Tom? Hell, they're all bunch 'o communists—whatta they care about business anyway?"

"With all due respect, sir, America could have been in orbit a year ago if we had an engineer in the White House instead of a politician."

"Maybe you should run next time." Kindelberger set down the last tchotchke and turned to Tom. "Two hundred years from now, some star-struck fifth grader will be sitting in physical-science class, and his teacher will tell him how Russia was the first country to place a satellite into orbit. It's an accomplishment that will live forever. No one will ever be able to take it from them."

"That's true, sir."

"Anyway, I had a long talk with von Braun. According to him, this propellant problem is the one bottleneck in the program. That girl of yours—Morgan—is the only thing keeping America from putting a satellite into orbit."

"What do you suggest, sir?"

"Any chance things would go a little faster if we put somebody else on it?"

"I honestly don't think so. And even if I found someone better, I wouldn't do it."

Kindelberger seemed taken aback by this response. He returned to the golf bag and removed a seven iron. "Why the hell not?"

"Because, sir; courtesy goes a long way."

This incorporation of his earlier advice put Kindelberger off-balance, and for once he had nothing to say. But Tom did.

"It's much more than that, sir. Fact is, I don't have anyone better. When it comes to theoretical performance calculations, Mary Sherman Morgan is not just the best in the company, she's probably the best in the world."

Kindelberger nodded. "I trust you, Tom. I know you would never do anything to jeopardize my company, or your job."

Kindelberger held the seven iron high above his head in what appeared to be a threatening gesture. Then he pulled it down, spread his feet in a drive position, then took a practice swing. He looked up, and on his face Tom could read the image of a golf ball sailing into the stratosphere.

As Kindelberger returned the club to Tom's golf bag, he asked, "You going to be at the Downey Golf Tournament this year?"

"Wouldn't miss it, sir."

Kindelberger nodded his approval, then pointed to Tom's desk.

"Ya know, you otta try to keep your desk a little cleaner."

The Dutchman was already out the door when Tom heard his boss say, "Tell Gloria I said hi."

Tom went to the window and watched the famous aircraft builder weave through the desk maze. This time there was no body-wake, as most of the employees had chosen the Kindelberger visit as an excuse to take an early lunch. Tom kept his eyes on the black suit and white hair until they disappeared through the front door into the lobby.

Then he called Mary's desk phone.

—/\/\/\—

The conference room normally held a maximum of twenty people. Today there was at least twice that many. Mary, Bill, and Toru were there, along with department heads and supervisors and two dozen engineers and technicians. Dieter Huzel, von Braun's right-hand man, was there as an NAA employee; he had been hired out of the Huntsville group two weeks before. The meeting had dragged on for more than two hours as they discussed and argued over the combustion instability problem with the A-7 engine burning hydyne. Several good ideas had been presented, and Tom Meyers decided it was time to close.

"We feel Mary's cocktail has what it takes to satisfy the army's propellant contract. But we can't tell von Braun to load it into the Redstone unless we can prove that it works. You've all got your assignments. Go and get me some good news."

23.

310 AT 1.75 AND 0.8615 FOR 155

"Engineers like to solve problems. If there are no problems readily available, they will create their own problems."

—SCOTT ADAMS,

WRITER AND ILLUSTRATOR OF THE *DILBERT* COMICS

A group of engineers and technicians had just completed a test of a small liquid-propellant rocket motor using LOX and pentaborane on VTS-2. There had been many delays, and the test-firing, originally scheduled for 11:00 a.m., did not come off until 8:40 p.m. Everyone was dog tired, and as soon as the all-clear signal was given, they all decided to go home. There would be enough time the following morning to examine the engine and look over the test results.

Two technicians, Mitch Hussleman and Bob Porvitz, stayed behind to unload a small, unused residue of pentaborane from the propellant tank.[1] Due to its high toxicity as a nerve agent, it was imperative that any unused pentaborane be disposed of without delay. Somehow during the process of transferring the fuel, a small amount leaked out and was inhaled by both men. Almost immediately, they lost all control of their arms and legs, and their bodies collapsed to the ground. They were fully conscious and alert, but could move very few of their muscles. They could not talk with each other, but both men understood right away what had happened.

It was after 10:00 p.m., and the test stand area was pitch dark. The men were alone, and no one was going to come save them. After a few minutes, Mitch discovered he could move enough muscles in his body that he could inchworm his way along the ground. He began to slowly make his way toward the blockhouse, sluggish and snaillike. If he could get to the blockhouse, a security guard might happen to drive by, and they would be spotted.

After a few minutes, Bob was able to mimic Mitch's inchworm technique, and he followed his colleague down the road.

—⌁—

When Mary arrived at the bowl for the seventh attempt at a full-up hot-fire test of the new propellant, she saw two ambulances parked at the blockhouse. One drove off just as she walked up. The second was preparing to leave. Mary approached Art Fischer, who was standing next to the blockhouse.

"What happened?"

"Bob and Mitch. We're not sure, but we think they may have toxic poisoning of some sort. They were working with pentaborane last night. We found them this morning lying next to the blockhouse."

"Are they alive?"

"So far. I gotta go."

Art jumped into the Jeep with three other technicians, and they followed after the second ambulance. Roger was standing nearby, and he turned to look at Mary.

"We're still going to fire your engine. Everybody here knows how important this project is."

Mary nodded. "Thank you."

Over the past three weeks, Mary and a team of engineers had been working on the A-7's instability problem nonstop. No one could really look inside of a large liquid-fuel rocket engine while it was running to see what was going on, and there were few measurement options available. Bill Vietinghoff had arranged to place pressure transducers inside one of the combustion chambers to take measurements, and a team of engineers including Bob Levine, Joe Friedman, Paul Castenholz, Bob Lawhead, Paul Kisicki and many others, analyzed those measurements.[2] Also working the problem were two of von Braun's colleagues, Walther Riedel and Dieter Huzel, both of whom were now employed by North American.[3]

There was no sound scientific theory that covered the instability problem, so the engineers had no choice but to make "monkey-cage guesses" —taking wild ideas, applying them, and seeing what worked. It was a very expensive example of trial and error. On the morning Hussleman and Porvitz were carried away in ambulances, the engineers had brought to Santa Susana a Redstone engine with a redesigned propellant injector. They had decided to make a subtle change in the way the fuel and oxidizer were fed into the chamber and mixed. Would it make any difference? No one knew—without a greater understanding of what was going on inside the 4,000-degree chamber, they had no choice but to experiment.

Two hours later, the countdown reached zero, and the Redstone A-7 engine, serial number NAA-110-39, ignited perfectly. Mary was not superstitious, but she found herself crossing her fingers as the burn approached the 40-second mark.

It kept going past 40. It kept going past 50 seconds . . . 60 seconds . . . 70 seconds . . . and it did not stop until it hit 155 seconds. Even before the all-clear siren blew, everyone in the blockhouse was shouting and screaming and celebrating.

A total of three successive tests had to be completed to comply with the contract, so after everyone calmed down, Roger stood on a chair to get their attention.

"Okay, people. Let's load it up. We need to do that two more times!"

By 10:00 that evening, the A-7 had undergone two more successful firings. Mary grabbed the phone and called Tom Meyers at home.

"You need to get that engine and a tank-car load of hydyne to Cape Canaveral, Tom. We're going into space."

—◦--\/\/\/--◦—

After three days in the hospital, both Mitch Hussleman and Bob Porvitz fully recovered from their ordeal and were back to work at the Santa Susana Field Laboratory. However, from that day forward, both men refused to work on any project that involved pentaborane ever again.

24

THE LAW OF UNINTENDED CONSEQUENCES

"Nobody, who has not been in the interior of a family, can say what the difficulties of any individual of that family may be."

—JANE AUSTEN

O ne day I got this crazy idea that I should write a book about my mother. The experience with the stage play had given me some detective skills, but I knew a book would have to be much more comprehensive than a ninety-minute play. I needed help; I needed more information. So in August 2010, I put up a blog, www.maryshermanmorgan .blogspot.com, to be a gathering place for my mother's friends, family, and former coworkers to weigh in with information, anecdotes, documents, and the like. The intentions of this were good, but it failed to work the way I intended. So many aspects of my mother's life are so deeply personal that few people want to post. In more than two years, it garners only twenty-two posts, and almost half of them are mine.

Yet the blog does yield one major unexpected discovery: My lifelong belief that I am my mother's first-born child turns out to be (wait for it, wait for it): wrong. For decades, a secret has been bubbling just beneath the surface of our middle-class suburban household: In 1944, Mary Sherman had a daughter out of wedlock and gave her up for adoption. I was fifty-seven years old when I discovered this for the very first time. Prior to that, not even a hint. It's like a family version of the Manhattan Project: People in the know know, and no one else does. And those few in the know have kept it a well-guarded family secret.

I will let you uncover this revelation the same way I did: through an exchange of e-mails.

From my sister Monica Weber (no relation to Bill Webber) on August 30, 2010:

Hi George:

The following is a message I got on Facebook from Dorothy Hegstad.[1] Make of it what you will. Coming from Aunt Elaine, I am dubious. And why would you tell someone this through an e-mail or Facebook?

Monica

Re: Mary Morgan

When my husband was doing the family tree thing with calendars, Aunt Elaine told him that your mom had been going to the college in Minot, ND, and the reason she left there was that she was about to become an unwed mother. She went to live with her Aunt Ida LaJoie Miller. Ida's daughter adopted the baby girl and Mary went on to college. I don't know if this is something that is common knowledge in your family or not. But that is one of the things that George put on *Wikipedia* that he didn't seem to know about.

Dorothy Hegstad

After I send an e-mail to Dorothy Hegstad for more information, I receive this reply:

Sept. 5, 2010

Hi, George, I don't know a lot about your Mom. They left ND shortly after I was born. My husband came out to LA in the mid-nineties and Aunt Elaine told him that Mary had been attending school (college) in Minot and became pregnant. She then went out to Ohio and stayed with her aunt Ida LaJoie Miller. When her daughter was born she was adopted by Ida's daughter, Ruth. Ruth had been married over 8 years at that time and apparently couldn't have children. My husband does a lot of family work but we've never looked into this. This is one of the reasons that Elaine picketed for anti-abortion. I remember my dad talking about this but it didn't register with me as I was only 9 or 10. It would be interesting to find out if you have an older sister. My dad sounded like he was proud of her for giving her baby to someone that could give her a good home.

Dorothy Sherman Hegstad

I send her this reply:

Sept. 5, 2010

Hi Dorothy:

Thanks for the reply. The college she attended—I assume that would be Minot State University. Right? Do you know what her major was, or how long she attended? After she moved to Ohio did she attend another college?

I know she worked in Ohio for a while. Do you know what company she worked for?

Of course, the big question is how do we get in touch with anyone who would know about the adoptive daughter. Did Ruth ever have or adopt any other children?

You can get some information on the play I wrote about Mary at: www.nevadabelle.com.

Thanks for your help!

George Morgan

A day later, I get this from Dorothy:

Sept. 6, 2010

In the family information we have Ruth Miller was married to Dudley Hibbard. We show 2 daughters. The first one was Angela Marie Hibbard, born Oct. 20, 1942. I believe this is your half sister. Another daughter was adopted in 1944 a year and a half later. Ruth was married 10 years before Angela appeared. In those days if a person was married they had children if they could have them. The timing is right for Mary having a baby in 1942. I believe she attended Minot college for two years and also went to college in Ohio. I have no idea where a person would go to get that information. Minot State University is what it's called now. I don't know what it went by then. I took some classes there in 1975 and it was called Minot State then. I wonder if there would be a way to contact with some of the LaJoie family to find out more. Leona LaJoie lives in Colorado Springs and I talk to her 2 or 3 times a year. I will see what she has to say. Since Ruth was born in 1910 I'm sure she is no longer with us.

Sincerely,

Dorothy

Based on that e-mail, it would seem I have an older sister and her name is Angela Hibbard. What I don't know yet is that this e-mail is riddled with errors. There is no hard evidence my mother ever attended Minot State, and she gave birth to a daughter in 1944, not 1942. Without checking the information, I put a post on my blog that the Morgan kids have an older sister and her name is Angela Hibbard.

Two weeks later, I receive this e-mail from someone I've never heard of:

Sept. 24, 2010

Dear Mr. Morgan,

Could you please contact me sometime about your mother. I may have some information about her that you may or may not already know.

Blessings!

Ruth E. Fichter

Included are two phone numbers she could be contacted at. Since I have no idea who "Ruth E. Fichter" is, I make the invitation to contact her a low priority. There is no Fichter anywhere in the family genealogy, what importance could it really have?

Weeks go by, and once in a while I think about the e-mail, so mysterious and cryptic. I fail to see this as the sign it was; a woman who led as mysterious a life as my mother would probably be the kind of person who would generate mysterious e-mails, even after her death. Still, I keep putting "Ruth E. Fichter" out of my mind. After almost a month goes by, I open the e-mail again and dial the number. It's early evening in California, around eight o'clock. I don't recognize the area code, so I have no idea where I'm dialing.

After about four rings, a sleepy voice answers.

"Hello?"

"Hi. May I speak to Ruth Fichter?"

"This is Ruth."

"You sound tired. Did I wake you up?"

"I wasn't asleep yet."

"Sorry. Where am I calling?"

"Detroit."

"Oh. I'm sorry. This is George Morgan. You sent me an e-mail."

"Yes, George."

"You said you had some information about my mother I might not already have."

"Yes. I have some information about your older sister."

"Right. There's a lady in Arizona named Dorothy Hegstad who claims our mother gave birth to a daughter many years ago, before she moved to California. Most of us out here are dubious about that story, but I'm looking into it."

"What have you found so far?"

"Not much. Dorothy gave me a name: Angela Hibbard. Whoever this person is I guess she's my sister."

"No. *I'm* your sister."

I'm standing in my backyard at that particular moment and almost fall over. What does a person say when, at the age of fifty-seven, they suddenly discover they have an older sister, and that her existence has been intentionally hidden? And what does that person say when they suddenly find themselves talking to that sister on the phone? The entire experience feels like a plot ripped from a *Lifetime* movie. We talk for a little while, then she excuses herself—time for bed. She tells me she had found me through the blog I had set up; some Sherman family member had told her about it. Ruth and I promise to stay in touch, and then we hang up.

So what do I tell my family? What do I tell my father? Or should I tell anyone at all? I walk back into the house, close the sliding-glass door, then spend the evening pretending to my wife and children that my life has not just changed dramatically.

A few days go by, and I summon the courage to call my father and tell him what I have discovered. I know it will shock him, perhaps hurt him. But it has been six years since my mother passed away, and I'm hoping time will help heal any wounds he might incur. When I call his phone, he picks up after about five rings.

"Hello?"

"Dad, it's me."

"Hey, George. How are things?"

We make some small talk for a few of minutes, and I ask him a couple of research questions about North American Aviation (a subject that always puts him a good mood). Then I hit him with it.

"Dad, you're not going to believe this, but I just found out mom had a daughter out of wedlock before she met you, and that she gave the baby up for adoption."

His answer is not what I expect.

"I know," he says. "But how did *you* find out."

I had created the blog with the intention of bringing to light accomplishments my mother made in the field of aerospace. Two years after first posting it, the blog yields almost none of the information I hoped to obtain. Yet were it not for the blog, I might never have discovered that all these years, I've been a little brother, and never suspected a thing.

A few months later, Ruth Fichter flies out to Southern California, and we have a rare Morgan family reunion. We take her to the beach (she had never seen the Pacific Ocean before), the Getty Museum, and the Griffith Observatory. We do lunch, we do dinner, we hang out. We talk, we laugh, we get to know each other. She spends a week with us, and it's as if we'd all known each other forever.

By the time she flies back to Detroit, I realize I'm not going to be able to write the book I intended.

25.

SATELLITE WITHOUT A NAME

**There is just one thing I can promise you about the
outer-space program: your tax dollar will go farther.**
—WERNHER VON BRAUN

Wernher von Braun sat alone in a darkened room. General
Medaris had ordered him to Washington, DC. In the unlikely
event the Redstone succeeded in launching America's first satellite on
its very first attempt, the army wanted its most public cheerleader to
be in the nation's capital. They would need a trusted public figure to
make the official announcement before the media, and who better for
that than Walt Disney's Dr. Space?

In a few hours, all of his life's work would be gambled on a single
rocket launch. It seemed as if every waking moment of von Braun's
earthly existence since age 12 had been prologue for this night, and
now he would not be allowed to watch the launch. There would be
no closed-circuit television or real-time voice link with the Cape.
Hundreds of miles away, someone's pulse would quicken as they
announced "Ignition!" and that person would not be him. When the
Redstone engines fired and rumbled, he would neither see nor hear
nor feel any of it. It was disconcerting. It was borderline humiliating.
It was unquestionably depressing. Wernher von Braun was a smoke-
and-fire man, but on the most important smoke-and-fire night of his
life, he was writing a PR speech in a dimly lit room 800 miles from the
launch site. There would be no phone calls from friendly colleagues;
for security reasons, the announcement of the success or failure of the
rocket would arrive in the form of an impersonal printout from a *rat-a-*

tat teletype machine installed in the next room. Teletype was as far as coded technology had progressed in January 1958.[1]

Wernher's pen stopped moving as his thoughts began to wander. There was an image that suddenly invaded his mind. He was seven years old and obsessed with building things. He wanted to build a tree house in the yard of his family's Berlin home. He had neither money nor materials, so he took a number of books an aunt had just purchased for him and attempted to return them at the bookstore for cash.[2] Before he could consummate the deal, his mother showed up and threw him and his books into the car.

The memory made Wernher smile.

I would certainly enjoy a nice tree house now.

Wernher's attention refocused, and his pen resumed writing.

—ᐰᐰᐰ—

A mere thirty miles away from North American Aviation, the Jet Propulsion Laboratory (JPL) in Pasadena had set up a room where a group of its engineers could gather to hear the results of *Explorer 1.* Worldwide satellite-listening stations would one day be set up, but in 1958, there were only three: Cape Canaveral, Huntsville, and JPL. As the westernmost station, JPL would be the first to have an opportunity to hear the satellite's signal, so it would be the first to know whether America had succeeded or failed in its latest satellite-launch attempt. Because of its close proximity to NAA, several engineers had arranged a real-time audio link between the two organizations; NAA would know what JPL knew, and at the same moment. Such a link-up was an easy arrangement for two organizations heavily loaded with electrical engineers and amateur radio licensees. As the time approached for launch, dozens of engineers filtered into the room, coffee mugs in one hand, cigarettes in the other.

—ᐰᐰᐰ—

The Redstone had been scheduled for launch on January 29, but high-altitude winds, at speeds of up to 175 knots, delayed the firing for two days.[3] They finally succeeded in getting a green light to launch at 10:30 on the night of January 31.

General Medaris watched from a crowded bunker 100 yards away. As the gantry pulled away from the rocket, Medaris would later describe what he saw:

"The missile stood like a great finger pointing to heaven—stark, white, and alone on its launching pad."[4]

There were a couple of minor delays, but finally at 10:48 p.m. General Medaris told the launch crew, "Firing Command."[5] A member of the crew pulled a pin that initiated the top-stage spinning (the poor man's guidance system before better methods were invented). Once the upper stage had reached its proper RPM—a wait of about thirteen seconds—the bottom-stage engines were fired. The massive rocket stood motionless for three seconds as the engines built up thrust. Once that thrust reached the point where it overcame the rocket's weight, the Redstone began a slow, upward movement. Its ascension seemed impatient at first, as if it wanted to take its time leaving Earth. But it continued to accelerate, gained altitude, climbed steadily, and before long became a mere speck in the nighttime Cape Canaveral sky.[6]

—◦∿∿∿◦—

The man assigned to track the satellite and determine its level of success or failure was Al Hibbs, an engineer from JPL on loan to the Cape. Despite the fact that it would be ninety minutes before they could be certain orbit had been achieved, General Medaris could not wait. Thirty minutes after launch, he approached Hibbs and demanded to know if the satellite was in orbit. Hibbs' reply—a classic display of engineering ass-protection—has become part of early Space Race lore.

"Hibbs! I need to know. Is the satellite up?"

"I can tell you with 95 percent confidence that there is a 60 percent chance the satellite made orbit."

"Don't give me any of this probability crap, Hibbs! Is it up or not!?"

"It's up."[7]

America had launched its first satellite and in so doing let the world know there was more than one competitor in the game.

—◦–WWv–◦—

I'm five years old and sitting in a chair in our garage in Reseda, watching my dad work on the forest-green VW with the small rear window. He's underneath the car, fiddling with something. I don't know what he does under that car, but he's always under it. He's the consummate grease monkey. Suddenly mom runs in. She's wearing an apron smudged with spaghetti sauce.

"Dick! It's up! The satellite. Von Braun just launched our country's first satellite. It's on the TV—right now."

My dad gets out from under the car and they both run into the house, leaving me alone.

What's a satellite? I ask myself. I get off the chair and follow them inside. It would be another twenty years before I was told why this event was so significant to my mother.

—◦–WWv–◦—

A reporter from the *Washington Post* raised his hand.

"What is the name of your satellite?"

Wernher von Braun was hosting his first post-launch news conference. With that one simple question, he realized what a mistake it had been to spread himself so thin—to micromanage so many different aspects of the Redstone's design, construction, and testing. So obsessed had he and everyone else been with the nuts and bolts of putting a satellite up that no one had given a moment's thought to naming the thing. Now he was being asked a question so unassuming and fundamental, yet he had no answer. At moments of historic significance, it was crucial for the right people to say the right things. Such comments would become part of history—they had to be thought out, not improvised.

The Russians were smart enough to name their satellite; why weren't we?

Wernher knew he should have learned this lesson by now, yet here he was reinventing the wheel. If he had been facing a roomful of engineers there would have been not a single question he would be unable to answer. But this was a media event, not a symposium.

What is the name of your satellite?

This was exactly the kind of question he should have anticipated. Naming things was one of the perquisites of the human race. Cats had names. Dogs had names. Even cars and dolls and breakfast cereals and rocket propellants had names. Certainly America's first satellite should have had a name. Assigning names to people and animals and objects and things both animate and inanimate was an integral part of being human. Science and humanity, could they coexist in harmony? Could they coexist at all?

To the reporter, Wernher gave the only answer possible:

"The satellite has not yet been given a name. It will have one shortly, and we will let the media know."

After the press conference, an army corporal drove Wernher back to his hotel. As he rode quietly in the back seat, Wernher pondered what kind of name the satellite should be given. He had a few ideas, and he would pass them along to General Medaris. It would, after all, be up to the army. That's what Wernher had failed to remind the reporters at the press conference; that the Redstone/Jupiter C launch vehicle and its satellite, like himself, were property of the US Army. It was a military project bathed in secrecy and paid for with a weapons budget. The Redstone rocket that had so dramatically lifted America's spirits and its first satellite was nothing more than an ICBM gussied up with a party dress and makeup. Like the German V-2, the Redstone was a weapon designed to kill. Yet in the morning, Americans across the country would be celebrating "their" rocket, "their" satellite, "their" success, as if the whole venture had been a civilian, rather than military, endeavor. Over the next few weeks, there would be speeches, parades, award ceremonies, talking-head correspondents, kudos, congratulatory backslapping, and endless parties. Like throwing a baby shower for a girl who had been gang-raped, the whole circus would turn a blind eye to what got them there in the first place.

I know why the satellite doesn't have a name. It doesn't have a name because an army funded it, and armies only give names to weapons.

The lights of the nation's capital twinkled near and far, but Dr. Wernher von Braun barely noticed them. He had helped America regain its *Sputnik*-besmirched reputation; he was a national hero. But heroism is almost always transitory. A soldier loses a leg in battle, gets a medal and a warm welcome home, then spends the rest of his life in the waiting room of some dingy VA hospital. For now, Dr. von Braun put aside thoughts of what to name America's satellite; there was a much more important question demanding to be heard.

What happens next?

It was a question he could not stop thinking about.

Fate had steered his life in so many unexpected ways. Fate: Wernher felt like a puppet at the end of its strings, yet the strings had been of his own making. Would he ever have the freedom to pursue the goals and dreams that had always been in his heart? Would Fate make way for Destiny?

Will I ever be allowed to get out of the goddamn weapons business?

A hotel valet opened his door, and Dr. Wernher von Braun, former Nazi weapons designer and new American hero, stepped onto the sidewalk. He shuddered; it was Washington, DC, in January, and he was still wearing Huntsville clothing. He opened his wallet and tipped the valet.

In his room that evening, he fired off a letter to North American Aviation. America's first satellite had only been possible due to the invention of hydyne. Without it, the United States would have continued to trail the Russians for months, if not years. The only thing he knew about its creation was that a woman at North American Aviation had cooked it up. It was a curious cocktail of two little-used chemicals, and it had done its job perfectly. He did not know her name, but he wanted to thank her. Von Braun took out pen and stationery and wrote, *Dear Unknown Lady.*

People, like satellites, sometimes go nameless.

Ten years later, as more information about *Sputnik* came to be known, von Braun would discover that Sergei Korolev and the Soviet Union, rushing to win a race that only they were in, hadn't thought of naming their satellite either. The name *Sputnik*, like *Explorer 1*, turned out to be a post-orbit appellation.

Three days passed, then the mail hit like a hurricane. Von Braun's office was deluged with sacks of mail from young boys and girls across the country and around the world. They wanted to know how they too could become rocket scientists and participate in the new world of satellites and space travel. Over the coming weeks and months, it became apparent that von Braun and his engineers had achieved something far greater than merely putting a satellite into orbit; they had inspired a future generation of engineers. Though it would not be well acknowledged, overshadowed as it was by the spectacle of launches to come, the real legacy of Sergei Korolev and Wernher von Braun was the inspiration they afforded a generation of youth to pursue careers they might otherwise never have considered.

—⋀⋀⋀—

Time marched on, and in June 1971, I graduated from Chaminade College Preparatory in Canoga Park. I graduated with honors, but far from the valedictorian status my mother had earned at the same point in her life. For four years, I had been an active member of all the nerd alliances: Amateur Radio Club, Electronics Club, and yes, the Rocketry Club. In the Radio Club, I earned a general class amateur radio license: WB6ZUV, which I keep active to this day (though I rarely get on the air). In the Rocketry Club, I frequently joined with eight or ten other students for weekends in the Mojave Desert, flying homemade zinc-and-sulfur solid-fuel rockets. I would usually invite my parents to come. My father would often make the trip, but my mother never did. Other than a week or two of summer vacation, she hated leaving the comfort of the dining room with its coffee, cigarettes, morning paper, and deck of cards.

In 1980, I decided I wanted to build and fly a liquid-propellant rocket. At that time, I still had almost no idea of the vast amount of experience my mother had in the subject, otherwise I would have sought out her help. My father was still working full-time as an engineer, and he gave me some encouragement and technical advice. I then called Dan Ruttle, a friend of mine who worked in the aerospace and rocketry field. Together we sketched out a design of an ultra-simple concentric tank rocket four inches in diameter, six feet long. It would use nitric acid as an oxidizer and furfuryl alcohol as a fuel (a combination that is hypergolic—these chemicals ignite by simply coming in contact with each other). As I would find out many years later, this design held many similarities to a project my mother worked on at North American: the NALAR air-to-air missile.

Eventually my brother Stephen signed onto the project. Design, construction, and testing took seven years (not a hobby for everyone), and in July 1987, the three of us flew the rocket in Nevada's Smoke Creek Desert, an alkali sink just west of the Black Rock Desert. It is a place so desolate that if you saw it, you would swear no human would ever go there for any reason. Yet a few years later, the annual ritual known as Burning Man would bring 40,000 people to it every summer for a week.

After my mother's death in 2004, as I began to sift through our family history and discover my mother's accomplishments in rocketry and aerospace, I couldn't help but wonder why she would not embrace her sons even when they were participating in an activity in which she had a personal interest.

It's a question I will probably never have a good answer for.

I've always considered open caskets to be morbid. Perhaps it's because I've seen too many of them. When I was a boy, it seemed as if every few months the Morgans or the Shermans or someone close to our family was having a funeral. No weddings—just funerals (to put this experience in perspective, the first wedding I ever attended was my own). So, by the time 2004 came along, I was at a place in my life where I had seen enough open caskets for one lifetime and decided to avoid them by avoiding funerals.

Yet, when the day of my mother's funeral arrived, something had changed. As I approached her open casket, I was glad to be able to see her one last time. I hadn't been the best son in the world—hadn't visited my parents as often as I could have, though I lived only an hour away. And so on that day, I stood by her side and lingered for a while. Just me and my mom. Then I said good-bye.

•—⋀⋀⋀—•

My den is filled with boxes stuffed with reference books, history books, notebooks, photographs, and video cards filled with talking heads. I stare at them, realizing they constitute almost ten years of my life. All those interviews, all that research, and I still don't know very much about my mother. She was a genius at everything she did, but at nothing else was she more skilled than her prankish, stealthy campaign of personal self-erasure. Her efforts at living a full life on the one hand, while plotting the expunction of her name from history on the other, required a level of mental perspicacity far beyond the average person. Fortunately, her attempt to make herself well regarded in life but anonymous in death ultimately failed; she touched too many people.

Something I've learned from this experience is that history is just like the future: uncertain. I can't help but wonder how many other Mary Sherman Morgans have lived throughout human history—people with accomplishments significant enough to be historic, but whose exploits historians failed to properly record. How many battles have been won because a soldier did something heroic, but was forgotten? How many inventions have been attributed to the wrong person? How many name-less shoulders did others stand on to grab glory? For the rest of my life,

I will look askance at everything I've learned in every history class I've taken. It's all suspect now. I have become better aware; the imperfections and flaws of recorded human history are laid bare, and they are legion.

I start gathering the boxes and carrying them out to my car. They'll go into our storage unit just outside of town. Perhaps I'll need them again someday. Or maybe some future historian will need them. One thing's for certain: We're going to protect the records this time.

WINGS OF THE CONDOR

T wo slide rules. Some paper and paper clips. A pair of scissors. A nail file. A couple of chemistry reference books. All of Mary's accumulated personal property did not fill even a single cardboard box. She took a second look through each drawer and checked for the tenth time that she had the congratulatory letter from Wernher von Braun. Once she gave up her security badge, not even the power of god would get her back inside this building. When Mary was sure she had everything packed, she took one last nostalgic look around her workspace. Bill and Toru's desks were empty, as they were conducting engine tests on the Hill. There were so many new hires, so many new engineers she had not yet met. She looked around for a friendly face—someone, anyone who would just say, "Good work. We'll miss you." But there was no one. They were all busy with new contracts, new propellants, new engine designs, new injector problems, new instability dilemmas. She longed to be a part of it all.

Mary picked up her box and started the long walk toward the exit.

Joe Friedman, recently promoted to senior engineer, was standing at his desk, talking with two other engineers. As Mary approached, Joe stopped mid-conversation, faced her direction, and began clapping. His two companions did the same. Other engineers in a fifty-foot radius stood up at their desks and joined in. As Mary continued her final walk, the circle of standing, applauding engineers grew larger and larger as it sought to fill every square inch of the hangar-size building. Many of the engineers climbed up onto their desks so they could see and be heard. Soon all the engineers were on their feet, clapping and cheering and whistling and roaring.

Mary was especially proud of one particular change that had occurred since her first day of work: at least a dozen of those applauding engineers were women.

She stopped at the steel exit door, turned, and gave a final wave, utterly overcome with emotion. She pushed through the door into the lobby. Tom Meyers was waiting for her. As the door closed behind her, the roar of the crowd faded out.

"Sounds like you were having a party in there."

"Against company rules, I assume."

"I wanted to be the last person you saw before you left. Congratulations. You helped make our country, and our company, proud. Everyone will miss you."

Mary noticed that both of the receptionists behind the front desk were crying.

"Thank you. I'll miss my coworkers, but I'll especially the work." Mary removed her security badge, but Tom refused to take it.

"You're going to need that for a few more hours." He handed her an SSFL pass. "Technically you still belong to me until five o'clock. I want you to head up to the Hill."

Tom placed a thin manila folder into her box.

"They're doing a test on an upgraded A-7. I want you to do the specific impulse data reduction. You can turn your badge in when you're done."

"I was going to take the bus home. Richard has the car. I have no way of getting up there."

Tom turned toward the large window. For the first time, Mary could see her husband—his red crew cut shining bright in the California sun—standing in the parking lot beside their forest-green VW with the small rear window. The engine compartment was open, and he was waxing poetic about some gear or strut or piston with another engineer.[1]

"He loves that air-cooled engine."

"We've given Richard a pass. He's going to take you up to the Hill."

Mary nodded. "Well, good-bye."

One of the receptionists rushed to open the door for her, and

Mary Sherman Morgan left the Canoga Park office of North American Aviation (recently renamed Rocketdyne), for the last time.

—/\/\/\—

When Richard and Mary entered the blockhouse, it was festooned with garland and crepe paper. The decorations were cheap, simple, slapped together. Whoever had purchased them had probably spent no more than two dollars on everything. Engineers had a well-deserved reputation for parsimony, and Mary was one of them. In the corner was a card table, and on it was a chocolate cake she was sure had been baked by one of the engineers' wives. It had that kind of look to it. A hand-painted cardboard poster on one wall had a felt pen scrawl: "GOOD-BYE, ROCKET GIRL." There was a photo glued to it of an A-7 engine test-firing, its fiery exhaust showing the now-familiar color scheme of the LOX/hydyne exhaust.

A dozen engineers and technicians were present, and they gave her a short ovational welcome. Very short—as in about two seconds.

Then everyone got back to business, which was fine with Mary. The Santa Susanna Field Laboratory was about to have a test-firing: frivolity, nostalgia, and going-away parties were not on the program. Besides, who wouldn't rather see a large liquid-rocket test-firing over chocolate cake any day?

Bill Webber appeared around a corner and shook her hand.

"It's not too late to change your mind."

She gently patted her baby bump. "One kid at home and another on the way. I have other work to do now."

Bill nodded, then glanced toward one of the blockhouse windows. The test stand could be seen, quiet and still, several hundred yards away.

"Stick around—it's gonna be one helluva blast."

Mary held up the file. "I'm doing the isp data reduction."

"Working on your last day; that is so like you." Bill pointed to one of the empty desks. "You can use that one." Bill smiled and returned to his station.

—◦–∿∿–◦—

The California condor had been lucky, having managed to scavenge a hearty meal off an early-morning roadkill on Santa Susana Pass Road. A speeding motorist had struck the coyote hard as it rounded a sharp turn, and now nature, abhorrent of waste, had turned the carcass into breakfast for other links in life's food chain.

The condor snipped off one last piece of red meat, hopped into the air, then flapped its nine-foot wingspan, ascending swiftly into the Southern California sky. With the summer sun moving steadily upward, the air temperature was rising, as were the currents. The condor happily rode those updrafts, giving its tired wings a needed respite. Still, it would be nice to just sit for a while—to rest from a journey that had taken it from the misty mountains of Big Sur to the rocky scrub of the Simi Valley hills. Its keen eyes surveyed the landscape below, but the vulture's options were few, as the hills in this part of its flight zone were covered with hot, dry grass and very few trees.

Over a low rise, the condor glided, and in a moment, it spotted a compound of dusty gray buildings. She circled the compound once, but could see no shade at all. It was getting hotter now, and the rising thermals were tickling the rows of long feathers that formed the back of each wing. With her featherless neck exposed to the sun, the condor was in danger of dehydration if she failed to find water, or at least some shade. She was about to yaw left in preparation for a turn southward when she spotted it—a large bell.

At least two or three times a year, the condor would make her way up the cool, breezy coastline to the Santa Barbara Mission. There was a favorite place she enjoyed—and always tried to return to—a high perch above the squealing children, shouting adults, and moving cars. There, a small archway provided a buffer from the ocean breezes, and hanging from that archway was a copper bell providing both shade and shelter. Now, as she maneuvered out of the updraft and drifted downward, one thing became clear—this new bell was much larger than the one at the mission.

She landed softly in the sun, then took a half dozen hopping steps to reach the quiet shade of the bell. She marveled at how much cooler it was, and made a mental note of all the nooks and crannies above her she could hide in if predators approached.

The condor folded her wings together, closed her eyes, and rested from its long journey.

—✳︎✳︎✳︎—

Mary was searching the office area for a Marchand or Friden calculator, but could find none.

"Richard, could you ask the technicians where the Fridens are?"

A few moments later, Richard returned with unexpected news.

"They sent them to storage. They're being replaced by some machine made by IBM called a computer. It's being installed right now over at the Canoga Park building."

"Then what am I supposed to use for the data analysis?"

Richard handed her a slide rule. "I guess you'll have to do it the old-fashioned way." Then he returned to the control room to assist with the firing.

Mary opened the file Tom had given her and read over the test parameters. Or at least she tried to. Normally Mary would have no trouble focusing on her work, but life was changing, and major disruptions (like retirement and pregnancy) have a way of throwing people off their rhythm. She thought about her son and baby-to-be, and all the attendant responsibilities of being a 1950s middle-American, white-bread mother. She had enjoyed her work in the aerospace business, but it was time to move on. Tomorrow she would be a full-time homemaker and mother.

Mary caught herself daydreaming, a highly unusual activity for her. She began performing a calculation with the slide rule, but only a few seconds later, there was another disruption, this one coming from the next room. The engineers and technicians had abandoned their usual businesslike demeanor and were running around, talking and shouting. Suddenly one of the technicians ran in.

"Hey, Mary! You're not gonna believe it. There's a large bird sitting underneath the rocket nozzle. Kanarek says it's a condor."

Mary stood up from her desk, grabbed a pair of binoculars from a drawer, and took a look through one of the small, four-inch-thick block-house windows.

"It just flew right in and landed."

Mary focused the lenses as she searched the firing area. There was the 1,000-ton concrete and steel test stand, below which was the 84,000-pound-thrust A-7 engine. She could even make out the rocket engine's serial number, NAA-110-43, affixed to the turbopump casing. Government specs still required all engines to be test-fired three times before being shipped out to the customer. Today would be the third and final such test of engine NAA-110-43. They were behind schedule, and the customer was unhappy.

"I don't see it," she said.

"Right below the nozzle. It's in the shade—hard to see."

"Oh yeah—there it is." Mary realized Irving was correct; she had seen enough pictures in enough magazines to know.

"That is definitely a California condor."

"How can you tell?"

"Bare neck, bald head, black feathers. And its size."

With dispassionate efficiency, the loudspeaker announced the continued countdown: "Two minutes to ignition."

Bill Webber entered with another set of binoculars. He took his place at the second window. Both of them could see the condor casually preen itself, not a care in the world.

"He's going to get a 4,000-degree ass-kick if he doesn't fly his avian butt outta there soon."

Mary lowered her lenses and looked at him. "How do you know it's a 'he'?" She did not wait for a reply. "We need to stop the test."

"Hey, you know the rules. These hills are full of wildlife. If we stopped the tests for every rodent, squirrel, mountain lion, and bird, we'd never fire an engine."

"Not just any bird—a California condor."

"A vulture. A scavenger."

"We need to make some noise or something; scare it off."

Bill shook his head. "The test stand's too far away. You could go outside and scream all day—it would never hear you."

In front of them was an array of hundreds of knobs, switches, buttons, and meters. Mary leaned on the control panel and thought.

Focus on the problem—how do we scare it off?

"Noise. We need to make a noise." Her eyes again passed over the control knobs and meters. One of them was labeled "OXYGEN TANK VENT."

"The oxygen tank."

"What about it?"

"The tank is mounted close to where the bird is standing. If we vent the tank, it will make a loud hissing sound. Could scare it off."

The loudspeaker continued the count: "One minute, thirty seconds."

"Vent the oxygen tank—this late in the count? That's totally contrary to procedure."

"Please, Bill."

"I ain't takin' that responsibility. You wanna vent the tank, you'll have to run it by Kanarek."

Mary left the control room, ran down the hall, and burst into her supervisor's office.

"Irving—we need to vent the oxygen tank."

Irving looked up from his desk. "What the hell for? We're less than two minutes from test."

"That bird—the California condor—it's taking a siesta under the bell nozzle."

"I know." Irving laughed. "He's in for a little surprise."

"The sound will scare it away."

"You don't know that."

"Please."

"No way. That's completely out of procedure. You want to vent the tanks, you'll have to take it to Mansfield."

Without another word, Mary ran out. She took three running steps to the stairs, and climbed them two steps at a time. The office of Ned Mansfield was right around the corner. He happened to be standing in the doorway, examining some papers on a clipboard.

"Mister Mansfield." She took several labored breaths—it had been a long time since she had done this much exercise. And, of course, all those Winstons didn't help.

The loudspeaker interrupted: "One minute to ignition."

"Yes, Mary."

"We have a problem."

"What is it?"

"There's a bird—a condor—sitting under the nozzle. It's right under the engine . . ."

"You mean a California condor!?"

She nodded, still breathing hard.

"Oh my god—I've never seen one."

Ned headed quickly down the stairs, and Mary followed. He continued talking as they walked. "How long has it been there?"

"Just."

They entered the control room. The word had gone out to the technicians and engineers—all of whom now crowded around the three small windows, struggling for a view and fighting over the few available binoculars.

Ned entered and shouted. "Outta the way!!"

Like the Red Sea, they parted. Ned grabbed a set of binoculars from the nearest technician and held them up, carefully adjusting the focus knob.

"Holy Mother of God."

"She's beautiful."

Ned turned to Mary. "How do you know it's a 'she'?"

"I—I don't. That's not the point . . ."

Throughout the small blockhouse room, Mary noticed the techies and engineers had formed into two groups—those who were concerned for the noble bird's welfare, and those who were joking about having "burnt vulture for lunch."

Ned set the binoculars down as the loudspeaker blared, "Thirty seconds to ignition."

"You know, I'll bet if we vented the oxygen tank a couple of times, we could scare it off."

Normally Mary would take this opportunity to tell Ned that the oxygen tank idea had been hers all along, but today there was no time for posturing. It was at crunch times like this that ego-massaging could be considered an acceptable, if unpleasant, solution.

"That's a brilliant idea, Ned."

But Kanarek interjected. "What if the vent valve doesn't reclose? It could be a disaster."

Ned stared out the window. "Irving is right. This is outside of normal procedure."

"We don't have any other option, Ned."

The loudspeaker: "Fifteen seconds. Fourteen, thirteen . . ."

Ned inhaled and exhaled. "What the hell. Let's try it."

Keeping her eyes on the test stand, Mary grabbed the knob labeled "OXYGEN TANK VENT," and gave it a one-quarter turn. From this distance, and behind the two-foot-thick concrete blockhouse walls, there was no way any of them could have heard the hissing of the vented oxygen vapors. But just as a falling tree in a forest does make a sound even when no one is around to hear it, so too did the vented oxygen.

But nothing happened—the condor stayed right where it was.

Richard waded through the crowd and put his arm around his wife. Everyone was quiet.

At that moment, the automatic pre-launch warning siren began its wail. The addition of the siren to the hissing tank apparently was enough. After a few more seconds, the room filled with enthusiastic cheers, mingled with some disappointed "ohs," as the giant bird kicked off the concrete, flapped its wings hard, and pulled itself slowly into the sky.

Ned shouted. "Close the tank!"

"Done," said Mary.

The condor was gaining altitude. Even so, Mary knew the danger was not over; there was still the shock wave.

"Four . . . three . . . two . . ."

A quiet beat—then suddenly an intense hurricane of fire, a tornado of flame, roared from the rocket's nozzle at a speed greater than sound. A moment passed, then the shock wave collided with the blockhouse, pounding it like a 9.0 earthquake. Everyone watched the test without much emotion—they had experienced this a hundred times before. It was all very routine—science going about its daily tasks.

Except for the bird, it was just another day at the office.

—◦◦◦◦—

The sudden hissing sound had bothered the condor, and for a few moments she considered taking flight. But her keen eyes saw no evidence of a predator, and so she had remained. But when that discordant wail began, well—that was too much. She kicked her legs and ascended above the cement, climbing skyward. She crooked her head slightly to left, then right. She needed to find water, and began to search her memory for spots she had found in the past. She climbed another hundred feet in elevation, and then some unseen force kicked her hard from behind. It must have been very much like what the coyote had felt when it had been hit earlier that morning by the speeding car—a solid impact that sent the bird spiraling out of control. A mere second later, there was a roar—a roar like nothing she had ever heard—pounding her tiny ears and hammering every cell in her body. The ground was coming up fast, and then she hit, bounced twice, and landed in some scrub grass next to a mound of cactus. She had fallen past the summit of a ridgeline, which helped to dampen the sound. Time passed, and she waited. The sound eventually stopped, and the dry hills returned to solitude.

A long breath, and the condor once again took flight.

As she climbed beyond the ridgeline far above the large bell and its concrete compound, the condor began to feel something strange—a feeling she had never had before. A new instinct was kicking in and demanding she find a safe place to build a nest. This puzzled her, though she could not deny the power of the urge.

But first she needed water. There was a place she had visited once before—a cool, breezy spot with a year-round spring in the mountains about twenty miles northwest. The water there would be sweet and plentiful, and there would be many rocky outcrops on which to build a safe home. The California condor flapped her left wing, dipped her right, and banked for a wide turn. She caught an energetic thermal and glided on.

AUTHOR'S NOTE

It took me several years to convince myself this book could be written. As memoirs go, it was full of potential bear traps and land mines. How does one write a memoir about a person who should have been famous, but wasn't? A person for whom there was no historical record to sift through. A person who did everything they could to bury their accomplishments and legacy, like pirates burying booty on some desert island, then burning the map. There came a day when I decided there was only one way a memoir like that could be written: as my own journey and adventure in search for that hidden treasure.

There is wide disagreement in the Sherman family over why Mary was not allowed to attend school for several years. In fact, they seem to disagree on a lot of family-history details. My mother often told the story of how the State of North Dakota gave her a horse so she could get across the river to attend school, yet one Sherman-family descendent claims there was no river to cross at all. That was only one of many muddy research details I had to wade through during this project.

The genre of this book is what is referred to in English writing courses as creative nonfiction. In creative nonfiction, the story is true but the writer must be "creative," that is, they must use dramatic license to invent some details in order to make the work readable. This story is true, but many details had to be invented. Unless otherwise cited, the dialogue between the characters, especially the NAA employees, is invented, though the situations during which that dialogue occurs were actual events. For example, the conversation between Mary, Bill Webber, and Toru Shimizu about simultaneous nonlinear equations really happened, but none of the dialogue was recorded at the time. I pieced together plausible dialogue based on my recollections of many conversations with Mr. Webber and Mr. Kanarek, and the memories and suggestions they expressed to me.

Many details of the story of hydyne, and the early space program in general, have been poorly recorded or not recorded at all. General Medaris was a real person, and his working relationship with Wernher von Braun, along with his responsibilities over the Redstone program, are fairly well documented. But the actual mechanics of how the propellant contract was presented to North American were not made available to me by NAA's current corporate descendent: Boeing. Therefore, the character of Colonel Wilkins (invented) and his meeting with Tom Meyers (the actual NAA engineering manager) are invented. The incident with the rocket punching through the Berlin police station was a real event, but the names of the police officers involved could not be found, so I had no choice but to invent both of them. The accident at Santa Susanna involving two technicians inhaling pentaborane and having to inchworm their way to the blockhouse really happened, but Boeing would not release their names, so I had to come up with a couple of names for the sake of the story. Several minor characters are based on real people, but their names have been changed, too (e.g., Nick Toby). The dialogue for Private John M. Galione was inspired by Mary Nahas's story about her father's exploits, *The Heroic Journey of Private Galione*, but there is no record of the actual conversations in the German wilderness. Also, the story of the California condor taking shade beneath the bell of a rocket during the final countdown to its test firing is true, but its place in time has been altered somewhat for the sake of storytelling.

I hope no one will judge this work harshly for such inventions. Without some creative liberties, the legacy and contributions of Mary Sherman Morgan, as well as the other NAA employees mentioned in this book, would have been lost to history forever.

ACKNOWLEDGMENTS

This is an imperfect book. It's imperfect because it relates a chapter of history that has not been well recorded. There are many gaps in the narrative that might be filled someday, and I look forward to that happening. Boeing could have been of more help, but, for the most part, they chose not to participate. Joel Landau at Rocketdyne's photo and video department stepped up to the plate and provided several much-needed archival photographs, for which I am very appreciative. Despite the many barriers, the main goal has been achieved: to perform research into the subject matter and create a book that, like my play from 2008, might jog more facts into the open at some future time.

I want to thank my professors at California State University, Channel Islands, and University of California, Riverside, for their contributions to my belated creative-writing degree, a degree I did not seriously pursue when I should have (something about turning 50 made me want to get out of my easy chair and do something). Each one of these instructors has contributed in some way toward making me a better writer, and *in toto* bringing me to a place where I could finally bring into the light of day my mother's long-buried accomplishments. These professors, in no particular order, are Julia Balén, Joan Peters, Sean Carswell, Bob Mayberry, Andrea Marzell, Jacquelyn Kilpatrick, Ray Singer, Renny Christopher, Julie Barmazel, Luda Popenhagen, and Greg Kamei from Cal State Channel Islands, and Joshua Malkin, Deanne Stillman, William Rabkin, and Charles Evered from the University of California, Riverside.

I would like to thank all of those who helped with the research: my brother, Stephen Morgan, and my sisters, Ruth Fichter, Monica

Weber, and Karen Newe. They helped by providing needed documents and photographs from family history, along with details of stories and events related to them by our mother and other family members. Monica and Karen pitched in to do some interviews when it became apparent I would not have time to do them all myself. Dorothy Sherman Hegstad provided information no one else seemed to have, including a Sherman family genealogy. A big thank-you to Duane Ashby for writing my mother's *Wikipedia* page.

Friends, colleagues, and former coworkers of my parents were very supportive with historical details: Dan Ruttle, Walter Unterberg, Joe Friedman, Don Jenkins, Bill Wagner, and Irving Kanarek. A special shout-out to Bill Vietinghoff, who pulled some strings and arranged for me to tour the Santa Susanna Field Laboratory just two weeks shy of the manuscript's due date.

The book would never have come about without the stage play, and so I would like to thank Shirley Marneus, Brian Brophy, and all my friends at TACIT for their many efforts and contributions. The book also owes its existence to my tireless agent, Deborah Ritchken of the Marsal Lyon Literary Agency.

Neither the play nor the book would exist were it not for three individuals: my wife, Lisa, who has been so supportive, my father, G. Richard Morgan, who regaled me endlessly with stories from his aerospace golden years, and Bill Webber, one of my mother's favorite and closest colleagues at North American Aviation. Their enthusiasm for this project never wavered.

NOTES

CHAPTER 1: THIS IS A STORY

1. Robert S. Kraemer, *Rocketdyne: Powering Humans into Space* (Reston, VA: AIAA, 2005). Quote is from the liner notes.
2. Walter Unterberg, interview with the author, August 10, 2004.

CHAPTER 2: PRAIRIE GIRL

1. First quoted by H. G. Wells, made famous by President Woodrow Wilson. *Wikipedia*, s.v. "The War to End War," http://en.wikipedia.org/wiki/The _war_to_end_all_wars.

CHAPTER 3: THE RAKETENFLUGPLATZ

1. Bob Ward, *Dr. Space: The Life of Wernher von Braun* (Annapolis, MD: Naval Institute Press, 2005), p. 12.
2. Michael J. Neufeld, *Von Braun: Dreamer of Space, Engineer of War* (New York: Vintage Books, 2007), p. 47.
3. Erik Bergaust, *Wernher von Braun* (New York: Cobb/Dunlop, 1976), p. 40.

CHAPTER 5: I HAVE NO IDEA WHAT YOU'RE TALKING ABOUT

1. Mary's 1974 employment application, courtesy of G. Richard Morgan.

CHAPTER 6: "MOTHER DOES NOT ABIDE PHOTOGRAPHY"

1. Michael J. Neufeld, *Von Braun: Dreamer of Space, Engineer of War* (New York: Vintage Books, 2007), p. 70.

2. Ibid. and Bob Ward, *Dr. Space: The Life of Wernher von Braun* (Annapolis, MD: Naval Institute Press, 2005), p. 19.

3. Unlike rockets, fighter aircraft were forbidden under the Versailles Treaty. Neufeld, *Von Braun*, p. 74.

4. Project Paperclip. Ward, *Dr. Space*, p. 59.

5. Neufeld, *Von Braun*, p. 71.

6. Ward, *Dr. Space*, p. 17.

7. Neufeld, *Von Braun*, pp. 74, 81.

CHAPTER 7: THE GREAT ESCAPE

1. Bob Ward, *Dr. Space: The Life of Wernher von Braun* (Annapolis, MD: Naval Institute Press, 2005), p. 17.

2. Michael J. Neufeld, *Von Braun: Dreamer of Space, Engineer of War* (New York: Vintage Press, 2007), pp. 82–85.

3. *Wikipedia*, s.v. "Kummersdorf," last modified May 9, 2012, http://en.wikipedia.org/wiki/Kummersdorf.

4. Neufeld, *Von Braun*, pp. 110–12.

5. Ibid.

6. James J. Harford, *Korolev: How One Man Masterminded the Soviet Drive to Beat America to the Moon* (New York: John Wiley & Sons, 1997), p. 49.

7. Ibid., p. 50.

8. Ibid.

9. "Gulag: Soviet Forced Labor Camps and the Struggle for Freedom," Center for History and New Media, George Mason University, http://gulaghistory.org/nps/onlineexhibit/stalin/work.php (accessed April 11, 2013).

10. Stanislaw J. Kowalski, "Kolyma: The Land of Gold and Death," ch. 7, http://www.aerobiologicalengineering.com/wxk116/sjk/kolyma7.htm (accessed April 30, 2013).

11. According to a 1974 job application (courtesy of G. Richard Morgan), Mary graduated from Ray High School on May 31, 1940.

12. George Richard Morgan and Elaine Sofio, interview with the author, March 23, 2002.

CHAPTER 8: A LITTLE OF THIS, A LITTLE OF THAT

1. Sister Mary Nadine Mathias, Sisters of Notre Dame, e-mail to the author, September 20, 2012.

2. Ibid.

CHAPTER 9: AN ODD NUMBER

1. Letter from Mary to the adoptive mother in 1944, six weeks after the birth of her daughter in Philadelphia. Ruth's birthdate is April 13, 1944; her adoptive parents, Dudley Irving Hibbard and Mary Grace Hibbard, were from Huron, Ohio. This information was obtained from a Probate Court paper "Adoption of Mary G. Sherman" signed by Judge John W. Baxter and dated June 8, 1946, in possession of author.

2. "JoanFarley1946," "I Understand Mother, I Just Want to Know What Happened," Adoption.com Adoption Forums, originally posted March 30, 2008, http://forums.adoption.com/making-contact-communicating/330379-i-understand -mother-i-just-want-know-what-happened.html (accessed April 19, 2013).

3. "Sherry Nelson," "Saint Vincent's Home," Ancestry.com Message Boards, originally posted December 17, 2002, http://boards.ancestry.com/thread .aspx?mv=flat&m=1940&p=localities.northam.usa.states.pennsylvania.counties .philadelphia (accessed April 19, 2013).

4. This is the only version of this letter the author has found. Mary Hibbard gave it to her adopted daughter, Ruth Hibbard, at some point. Ruth eventually married a Mr. Fichter, and she kept the letter all these years. When I came out of the woodwork, Ruth sent me a thick folder of family documents, including a copy of the letter.

5. St. Vincent's allowed unwed birth mothers to pay for some of their expenses by working in the hospital for a period of time after giving birth.

6. Apparently, Mary was expecting a visit from her brother Vernon Sherman. This is odd since she was in Philadelphia at the time, and he, pre-sumably, was still in North Dakota. However, since the war was still on, Vernon may have been stationed at a military base nearby.

CHAPTER 10: HIDDEN FORTRESS

1. Mary Nahas, *The Heroic Journey of Private Galione* (North Carolina: Mary's Designs, 2012), p. 53.

2. Ibid.

3. "World War 2 Weapons: German Machine Guns; MG-42," http://ww2 weapons1.tripod.com/worldwar2/id15.html.

4. Erik Bergaust, *Wernher von Braun* (New York: Cobb/Dunlop, 1976), pp. 89–91.

5. Ibid., p. 91. Bergaust places the number of relocated German engineers at 500. However, in his book *The Nazi Rocketeers* (Mechanicsburg, PA: Stackpole Books, 2007), Dennis Piszkiewicz sets the number at 400 (see pp. 190, 191).

6. G. Richard Morgan, interview with the author, April 30, 2000. Dieter Huzel worked a short time at Fort Bliss and White Sands before obtaining a more permanent position with Rocketdyne in Canoga Park. He, Mary, and her husband Richard worked together at times.

7. Bergaust, *Wernher von Braun*, pp. 90, 91.

8. Ibid., p. 91.

9. Mark D. Bowles and Robert S. Arrighi, "NASA's Nuclear Frontier— The Plum Brook Reactor Facility," published August 2004, p. 29, http://history .nasa.gov/SP-4533/Plum%20Brook%20Complete.pdf.

10. Ibid., p. 29.

11. Ibid., p. 30.

12. Ibid., p. 31.

13. Ibid., pp. 9, 10.

14. Ibid., pp. 22–24.

15. Ibid., p. 27.

16. Ibid.

17. Nahas, *Journey of Private Galione*, p. 59.

18. Ibid.

19. Ibid., pp. 63–67, 78.

20. Robert S. Kraemer, *Rocketdyne: Powering Humans into Space* (Reston, VA: AIAA, 2005), p. 11.

21. Piszkiewicz, *Nazi Rocketeers*, p. 191.

22. Bergaust, *Wernher von Braun*, pp. 90, 91.

23. Ibid., p. 91.

24. Piszkiewicz, *Nazi Rocketeers*, pp. 190, 191.

25. Ibid., p. 193.

26. Michael J. Neufeld, *Von Braun: Dreamer of Space, Engineer of War* (New York: Vintage Press, 2007), p. 197.

27. Ibid., p. 198.

28. Ibid., p. 199.

29. Ibid.

30. Ibid., p. 200.

31. Ibid.

CHAPTER 11: A NEW KIND OF WAR

1. G. Richard Morgan, interview with the author, December 5, 2005.

2. Irving Kanarek, interview with the author, Costa Mesa, March 6, 2011.

3. Two years after Mary stood at that intersection, the airport would be renamed Los Angeles International Airport.

4. Irving Kanarek, interview with the author, Costa Mesa, March 6, 2011. Though everyone, including Kanarek, agrees that he was fired from North American, there are several differing stories on what triggered his termination. Two engineers I interviewed said that Kanarek was fired after inadvertently leaving a briefcase full of secret documents in a bar in Los Angeles. Kanarek insists that story has no truth. Kanarek's version of the secretary's "correction" was verified by other sources, including Bill Webber. However, after North American and the FBI conducted an investigation and discovered the secretary did indeed alter Kanarek's application entry, he was not offered reemployment, lending credence to the idea that there might have been other problems.

CHAPTER 12: WHITEWASHED IN WHITE SANDS

1. Bob Ward, *Dr. Space: The Life of Wernher von Braun* (Annapolis, MD: Naval Institute Press, 2005), p. 59.

2. Erik Bergaust, *Wernher von Braun* (Washington, DC: National Space Institute, 1976), p. 83.

3. Ward, *Dr. Space*, p. 21.

4. Michael J. Neufeld, *Von Braun: Dreamer of Space, Engineer of War* (New York: Vintage Books, 2007), p. 218.

5. Ward, *Dr. Space*, p. 59.

CHAPTER 13: ALIAS CHIEF DESIGNER

1. James J. Harford, *Korolev: How One Man Masterminded the Soviet Drive to Beat America to the Moon* (New York: John Wiley & Sons, 1997), pp. 57–63.

2. Matthew Brzezinski, *Red Moon Rising: Sputnik and the Hidden Rivalries That Ignited the Space Age* (New York: Times Books, 2007), p. 100. Brzezinski describes the living conditions at Tyuratam as "monastic."

3. Victor L. Mote, "Steppe," *Encyclopedia of Russian History*, 2004, available at Encyclopedia.com, http://www.encyclopedia.com/topic/steppe.aspx.

4. Brzezinski, *Red Moon Rising*, p. 107.

5. Ibid., p. 108.

6. Ibid.

CHAPTER 14: RED

1. Carla Rivera, "Caltech Named Best Research University in the World—Again," *Los Angeles Times*, October 5, 2012, http://latimesblogs.latimes.com/lanow/2012/10/caltech-tops-list-of-worlds-universities-again.html.

2. "Liquid Oxygen and Liquid Hydrogen Storage," NASA, last modified November 23, 2007, http://www.nasa.gov/mission_pages/shuttle/launch/LOX-LH2-storage.html.

3. *Wikipedia*, s.v. "Hydrazine," http://en.wikipedia.org/wiki/Hydrazine.

4. "Military: V-2 Rocket," Wikia, http://military.wikia.com/wiki/V-2_Rocket.

5. John D. Clark, *Ignition! An Informal History of Liquid Propellants* (New Brunswick, NJ: Rutgers University Press, 1972), p. 86..

CHAPTER 15: POLITICS, PHILOSOPHY, TELEVISION, AND *CUSH' SOBASH'YA*

1. Constance Green and Milton Lomask, *Project Vanguard: The NASA History* (Mineola, NY: Dover Publications, 2009), p. 180.

2. Mike Wright, "The Disney–Von Braun Collaboration and Its Influence on Space Exploration," NASA, MSFC History Office, http://history.msfc.nasa.gov/vonbraun/disney_article.html. Originally presented by the author at "Inner Space/Outer Space: Humanities, Technology and the Postmodern World," a Southern Humanities Conference in 1993; later included in Daniel Schenker,

Craig Hanks, and Susan Kray, eds., *Selected Papers from the 1993 Southern Humanities Conference* (Huntsville, AL: Southern Humanities Press).

3. Ibid.

4. Ibid.

5. As Brzezinski put it, "von Braun quickly became America's space prophet." Matthew Brzezinski, *Red Moon Rising: Sputnik and the Hidden Rivalries That Ignited the Space Age* (New York: Times Books, 2007), p. 91.

CHAPTER 16: YOUR VERY BEST MAN

1. The LOX/alcohol isp (specific impulse) value = 284 seconds, "LOX/Alcohol," *Encyclopedia Astronautica*, http://www.astronautix.com/props/loxohol.htm.

2. George P. Sutton, *History of Liquid Propellant Rocket Engines* (Reston, VA: AIAA, 2005), pp. 39–40, includes a brief description of Irving Kanarek and the invention of inhibited red fuming nitric acid.

3. G. Richard Morgan and Bill Webber, interviews with the author.

CHAPTER 17: WELCOME TO THE MONKEY CAGE

1. The "Talk" section of Mary's *Wikipedia* page can be found at: http://en.wikipedia.org/wiki/Talk:Mary_Sherman_Morgan.

2. Bill Webber, interview with the author, October 23, 2012.

3. *Wikipedia*, "Will Beback Banned," last modified July 16, 2012, http://en.wikipedia.org/wiki/Wikipedia:Arbitration/Requests/Case/TimidGuy_ban_appeal#Will_Beback:_banned (accessed April 16, 2013).

4. Robert S. Kraemer, *Rocketdyne: Powering Humans into Space* (Reston, VA: AIAA, 2005), p. 44.

5. This George is a reference, of course, to the author.

CHAPTER 18: THE MYSTERIOUS UNKNOWN PROPELLANT PROJECT

1. Michael J. Neufeld, *Von Braun: Dreamer of Space, Engineer of War* (New York: Vintage Books, 2007), p. 290.

2. Matthew Brzezinski, *Red Moon Rising: Sputnik and the Hidden Rivalries That Ignited the Space Age* (New York: Times Books, 2007), pp. 38, 39.

3. Ibid., p. 98.

4. Ibid., pp. 113, 114.

5. Pronounced "DEE-tuh."

CHAPTER 19: SMOKE AND FIRE

1. Bill Vietinghoff, SSFL tour guide spiel, November 10, 2012.

2. *Wikipedia*, s.v. "Petroleum," http://en.wikipedia.org/wiki/Petroleum.

3. *Wikipedia*, s.v. "Pico Canyon Oilfield," http://en.wikipedia.org/wiki/Pico_Canyon_Oilfield.

4. Irving Kanarek, Bill Webber, and G. Richard Morgan, interviews with the author.

5. *Wikipedia*, s.v. "Santa Susanna Field Laboratory," http://en.wikipedia.org/wiki/Santa_Susana_Field_Laboratory.

6. Bill Webber, interview with the author, October 28, 2012.

7. George Richard Morgan, interview with the author, September 23, 2012.

8. George Richard Morgan confirmed in his interview with the author that with some exceptions, engineers were rarely invited to firings and that a separate crew handled all test firings. This policy was verified by Bill Webber in his interview with the author.

9. Paul Costa, SSFL tour guide spiel, November 10, 2012.

10. Webber, interview.

11. Information in this section obtained from Webber and Morgan interviews.

12. These were called "SSFL oxygen races." Webber interview.

13. The firing sequence in this section is based the author's interview with Bill Webber on December 15, 2007.

CHAPTER 20: DON'T DRINK THE ROCKET FUEL

1. Interview with the author.

2. Background information for this chapter was supplied by Irving Kanarek and Bill Webber in interviews with the author from October 18, 2011, and December 15, 2007, respectively. The firing-sequence announcements were provided by Bill Webber.

3. The isp of hydyne is 309. "LOX/Hydyne," *Encyclopedia Astronautica*, http://www.astronautix.com/props/loxydyne.htm.

4. Robert S. Kraemer, *Rocketdyne: Powering Humans into Space* (Reston, VA: AIAA, 2006), p. 39. The 110-second burn time was so built into the design that the Redstone engines were given a numbering system based on it. The first engine built was named NAA 75-110-01. The "75" stood for the thrust: 75,000 pounds. The thrust was later upgraded to 84,000 pounds, but the numbering system remained unchanged.

CHAPTER 21: *PUSK!*

1. Matthew Brzezinski, *Red Moon Rising: Sputnik and the Hidden Rivalries That Ignited the Space Age* (New York: Times Books, 2007), p. 145.

2. James J. Harford, *Korolev: How One Man Masterminded the Soviet Drive to Beat America to the Moon* (New York: John Wiley & Sons, 1997), p. 129; and Brzezinski, *Red Moon Rising*, pp. 152, 153.

3. Brzezinski, *Red Moon Rising*, p. 153.

4. Ibid., p. 156.

5. Ibid. for Soviet firing sequence.

CHAPTER 22: THE DUTCHMAN COMETH

1. Matthew Brzezinski, *Red Moon Rising: Sputnik and the Hidden Rivalries That Ignited the Space Age* (New York: Times Books, 2007), p. 100.

2. Matt Bille and Erika Lishock, *The First Space Race* (College Station: Texas A&M University Press, 2004), p. 106.

3. Ibid., endnote 32 of chapter 6.

4. Constance Green and Milton Lomask, *Project Vanguard: The NASA History* (Mineola, NY: Dover Publications, 2009), p. 186.

5. Bille and Lishock, *First Space Race*, p. 109.

6. Michael J. Neufeld, *Von Braun: Dreamer of Space, Engineer of War* (New York: Vintage Books, 2007), p. 311.

7. Ibid., p. 312.

8. Ibid.

9. Ibid., p. 313.

10. Green and Lomask, *Project Vanguard*, p. 206.

11. Ibid., p. 208.

12. Ibid., p. 209.

13. "James Howard "Dutch" Kindelberger," Boeing: History, http://www.boeing.com/history/bna/biog.html.

14. *Wikipedia*, s.v. "United States Army Air Forces," http://en.wikipedia.org/wiki/United_States_Army_Air_Forces.

15. Bill Webber, interview with the author.

16. See the author's note.

17. Per Bill Webber, interview with the author, October 28, 2012, everyone at NAA was aware of Kindelberger's propensity to fire 5 percent of the employees every year.

CHAPTER 23: 310 AT 1.75 AND 0.8615 FOR 155

1. Bill Webber, interview with the author; see also author's note.

2. Bill Vietinghoff, interview with the author, November 30, 2012, detailing his contribution.

3. Robert S. Kraemer, *Rocketdyne: Powering Humans into Space* (Reston, VA: AIAA, 2006), p. 38.

CHAPTER 24: THE LAW OF UNINTENDED CONSEQUENCES

1. Dorothy Hegstad is a member of the Sherman family living in Phoenix, Arizona.

CHAPTER 25: SATELLITE WITHOUT A NAME

1. Michael Neufeld, *Von Braun: Dreamer of Space, Engineer of War* (New York: Vintage Books, 2007), p. 320.

2. Ibid., p. 18.

3. Matt Bille and Erika Lishock, *The First Space Race* (College Station: Texas A&M University Press, 2004), p. 128, and JPL *Explorer 1* archives, p. 20.

4. Bille and Lishock, *First Space Race*, p. 129.

5. Ibid., p. 130.

6. Ibid.

7. Ibid., p. 132.

CHAPTER 26: WINGS OF THE CONDOR

1. Information in this section gathered from G. Richard Morgan, interviews with the author.

INDEX

alcohol. *See* ethyl alcohol (as rocket fuel)